Berliner ökophysiologische

und phytomedizinische Schriften

I0037883

Hrsg. von Christian Ulrichs und Carmen Büttner

Lebenswissenschaftliche Fakultät,

Humboldt-Universität zu Berlin

Band 47

Hrsg. von

Dr. Dietrich Stephan

Julius-Kühn-Institut, Darmstadt

und

Prof. Dr. Carmen Büttner

Humboldt-Universität zu Berlin

Biological agents formulation and mode of application against strawberry diseases

DISSERTATION

zur Erlangung des akademischen Grades
Doctor rerum agriculturarum
(Dr. rer. agr.)

eingereicht an der
Lebenswissenschaftlichen Fakultät der Humboldt-Universität zu Berlin

von

Diploma in Lebensmittelwissenschaften und Technologien
Isabella Linda, Bisutti
geb. am 23.03.1973, Spilimbergo (Italien)

Präsidentin
der Humboldt-Universität zu Berlin
Prof. Dr.-Ing. Dr. Sabine Kunst

Dekan der Lebenswissenschaftlichen Fakultät
Prof. Dr. Bernhard Grimm

Gutachterin/Gutachter
1. Prof. Dr. Carmen Büttner
2. Prof. Dr. Johannes Jehle
3. ....

Tag der mündlichen Prüfung: 23.11.2018

Bibliografische Information der Deutschen Nationalbibliothek

Die Deutsche Nationalbibliothek verzeichnet diese Publikation in der Deutschen Nationalbibliografie; detaillierte bibliographische Daten sind im Internet über http://dnb.d-nb.de abrufbar.

1. Aufl. - Göttingen: Cuvillier, 2019
    Zugl.: Berlin, Humboldt-Univ., Diss., 2018

© CUVILLIER VERLAG, Göttingen 2019
    Nonnenstieg 8, 37075 Göttingen
    Telefon: 0551-54724-0
    Telefax: 0551-54724-21
    www.cuvillier.de

1. Auflage, 2019
Gedruckt auf umweltfreundlichem, säurefreiem Papier aus nachhaltiger Forstwirtschaft.

    ISBN 978-3-7369-7032-8
    eISBN 978-3-7369-6032-9

"When you have excluded the impossible, whatever remains, however improbable, must be the truth."

The Adventure of the Beryl Coronet

"There is nothing more deceptive than an obvious fact."

The Boscombe Valley Mystery

Sherlock Holmes

Table of contents

LIST OF ABBREVIATIONS AND SYMBOLS

1:100	one to hundred ratio
%	percent
±	plus and minus
$\times g$	gravity, measured in meters per second
°C	degrees Celsius
μg	microgram
ANOVA	analysis of variance
BCA	biological control agent/s
CFU	colony forming unit
cm	centimetre
cv.	cultivar
g	gram
h	hour
IPM	integrated pest management
KB	King's broth
L / l	litre
m	meter
mbar	millibar
mg	milligram
min	minute
ml	millilitre
MPN	most probable number
MS	microsclerotia
NL	normal litre
OD	optical density
Pf153	*Pseudomonas fluorescens* Pf153
PGPR	plant growth-promoting rhizobacteria
pH	potential of hydrogen
PPP	plant protection product/s
rpm	revolutions per minute
s	second
sp.	species
spp.	species pluralis
ssp.	subspecies
v/v	volume per volume
v/w	volume per weight
w/v	weight per volume

LIST OF TABLES

LIST OF FIGURES

ABSTRACT

Biopesticides are an interesting solution in integrated pest management (IPM). Developing a microbial antagonist to a usable product is a long path with many production, formulation and legislative hurdles. During the production and formulation steps, the efficacy against pests and pathogens, storability and ease of application have to be kept in mind to deliver a product that is accepted by the end consumer.

In this report the quality of the bacterium *Pseudomonas fluorescens* Pf153 (Pf153) was optimized by developing appropriate production technologies. Freeze-drying was chosen as conservation process and it was shown that the freezing rate and the added cryo-protectant agent (CPA) influence the survival of *Pseudomonas* spp. CPA can increase the viability during the process and during storage and influences the efficacy of the formulated bacterium but also the pathogen activity. In particular for Pf153 the survival after freeze-drying and storage could be increased by using 0.04-0.12 °C/min as freezing-rate, 30 °C for the drying process and lactose as CPA. Since production and formulation are closely linked, the fermentation parameters have also to be optimized for high biomass production with high survival during formulation, storage and with a high efficacy. The tested media, fermentation temperatures, mild heat shocks and pH changes influenced the viability of Pf153 after freeze-drying with survival rates varying between 28% and 100%. Efficacy against *Botrytis cinerea* on *Vicia faba* leaves was increased only when the bacterium was fermented at 20 °C. This temperature "stabilized" the survival after freeze-drying, increased the storage viability at 25 °C for 12 weeks and showed a slightly better efficacy against five diverse *B. cinerea* strains in dual culture tests.

Different behaviour under diverse conditions is a big issue in the performance of formulated bioproducts. Trichostar®, RhizoVital® 42 fl. and *Metarhizium brunneum* Ma43 (Ma43) were tested, in greenhouse and field trials, on various growth parameters and yield of strawberries in the presence of soil pathogens. In greenhouse trials, Trichostar® had a poor performance compared to the control sample without treatment, when the strawberry plants were inoculated with *Phytophthora cactorum* and *Verticillium dahliae*. On the other hand, it increased the studied parameters when no pathogen was inoculated. RhizoVital® 42 fl. showed a better performance than Trichostar® in the presence of the pathogens in greenhouse trials. In commercial fields, the two selected products showed a good performance with an increased strawberry yield of about 8% for Trichostar® and 6% for RhizoVital® 42 fl., in two consecutive years on two different fields. The soil properties influenced the performance more than the weather conditions, however the application schedule should be ameliorated. Ma43 was tested in greenhouse and field controlled trials. In foregoing trials it increased strawberry yield, but in the trials presented here, its positive influence could not be confirmed. The contemporaneous application of the Trichostar® and RhizoVital® 42 fl. with or without Ma43 did not show an increase in their efficacy. The compatibility of the micro-organisms with chemical pesticides (12 fungicides, one insecticide and one acaricide used in strawberry production) showed their potential in IPM.

The results presented here, show the potential of biopesticides and their dependence from production and formulation parameters. Storage and efficacy are influenced by biotic and abiotic factors and not all of them can be controlled. Field application tests with the end product are urgently needed, because until now lab results could not predict the performance at its application site. The choice of a bioproduct has to be rational and judicious since more selection parameters have to be kept in mind in reference to chemical ones. Bioproducts can help to increase production, as shown for strawberries, with less depletion and pollution and without great changes for the grower.

ZUSAMMENFASSUNG

Biopestizide sind im integrierten Pflanzenschutz (IPM) eine interessante Lösung. Die Entwicklung eines mikrobiellen Antagonisten zu einem verwendbaren Produkt ist ein viele Disziplinen umfassender Prozess. Die Produktions- und Formulierungsschritte müssen erforscht werden, die Wirksamkeit gegen Krankheitserreger und Schädlinge und die Lagerfähigkeit und einfache Anwendung der Produkte müssen erarbeitet und erhöht werden.

In diesem Bericht wurde die Qualität des Bakteriums *Pseudomonas fluorescens* Pf153 (Pf153) durch die Entwicklung geeigneter Produktionstechnologien optimiert. Die Gefriertrocknung wurde als Trocknungsmethode gewählt und diverse *Pseudomonas* spp. zeigten unterschiedliche Überlebensraten, bei unterschiedlichen Gefrierraten aber nicht so bei den Trocknungstemperaturen die weniger stammabhängig waren. Die Lebensfähigkeit während des Trockungsprozesses und der Lagerung wurde mit der Zugabe von Kryoschutzmitteln (CPA) verbessert. Das Überleben nach Gefriertrocknung und Lagerung konnte durch die Verwendung der angepassten Gefrierrate 0.04-0.12 °C/min, die Trocknung bei 30 °C und Laktose als CPA für Pf153 erhöht werden. Die Wirksamkeit der gefriergetrockneten Zellen wurde in diversen Biotests gezeigt und diese war mit der Wirksamkeit von frischen Zellen vergleichbar. Diverse Fermentationsparameter wurden getestet um die Lebens- und Lagerfähigkeit und die Wirksamkeit von Pf153 nach Gefriertrocknung zu erhöhen: unterschiedliche Medien, Fermentationstemperaturen, milde Hitzeschocks und pH-Änderungen. Die Überlebensraten variierten zwischen 28% und 100%. Die Wirksamkeit wurde durch Änderungen der Fermentationstemperatur erhöht, jedoch konnte der Einfluss des Zellalters auf die zunehmende Wirksamkeit nicht ausgeschlossen werden. Niedrigere Temperaturen (20 °C) erhöhten die Lagerfähigkeit bei 25 °C für 12 Wochen und zeigten die beste Wirkung gegen fünf verschiedene *B. cinerea*-Stämme in Doppelkultur-Tests gegenüber Zellen, die bei 28 °C gezüchtet und in derselben Phase geerntet wurden.

Unterschiedliche Wirksamkeit unter verschiedenen Bedingungen ist ein großes Problem bei der Leistungsfähigkeit der formulierten Produkte. Trichostar®, RhizoVital® 42 fl. und *Metarhizium brunneum* Ma43 (Ma43) wurden in Gewächshaus- und Feldversuchen auf verschiedene Wachstumsparameter getestet und der Ertrag von Erdbeeren in Gegenwart von Krankheitserregern ausgewertet. In. Gewächshausversuchen zeigte das Trichostar® im Vergleich zur unbehandelten Kontrollprobe eine schlechtere Leistung;

wenn die Erdbeerpflanzen mit *Phytophthora cactorum* und *Verticillium dahliae* beimpft wurden. Andererseits erhöhte Trichostar[R] die untersuchten Parameter, wenn kein Erreger geimpft wurde. RhizoVital[R] 42 fl., zeigte eine bessere Leistung als Trichostar[R] in Anwesenheit der Pathogene in Gewächshausversuchen. Einen erhöhten Erdbeerertrag von etwa 8% für Trichostar[R] und 6% für RhizoVital[R] 42 fl., wurde im kommerziellen Bereich, in zwei aufeinander folgenden Jahren auf zwei verschiedenen Feldern erzielt. Es scheint, dass die Bodenbeschaffenheit die Leistung stärker beeinflusste als die Witterungsbedingungen. Ma43 wurde ebenfalls getestet. In früheren Versuchen konnte der Erdbeerertrag erhört werden, aber sein positiver Einfluss konnte hier nicht bestätigt werden. Die gleichzeitige Anwendung aller drei Bioprodukte zeigte keine Steigerung deren Wirksamkeit. Die Verträglichkeit der Mikroorganismen mit im Erdbeeranbau gängigen chemischen Pflanzenschutzmitteln (12 Fungizide, ein Insektizid und ein Akarizid) zeigt ihr Anwendungspotenzial im IPM.

Die vorgestellten Ergebnisse zeigen das Potenzial von Biopestiziden und ihre Abhängigkeit von Produktions- und Formulierungsparametern. Lagerung und Wirksamkeit werden durch biotische und abiotische Faktoren beeinflusst die nicht alle kontrolliert werden können. Feldanwendungstests mit dem Endprodukt sind dringend erforderlich, da die Laborergebnisse bisher die Leistung am Einsatzort nicht vorhersagen können. Die Wahl eines Bioprodukts muss rational und vernünftig sein, und es müssen mehr Auswahlparameter berücksichtigt werden als bei chemischen Produkten. Bioprodukte können dazu beitragen, die Produktion zu steigern, wie es bei den Erdbeeren der Fall war, mit weniger Umweltverschmutzung ohne dass der Landwirt große Veränderungen vornehmen muss.

CHAPTER 1

GENERAL INTRODUCTION

Bisutti Isabella L.

New challenges and incentives in agriculture

Modern farming is confronting an increasing demand for steady supply and high-quality fresh goods (Berninger et al., 2018; Emmert and Handelsman, 1999; Gamliel, 2010). Nowadays, the way to make and provide healthy products is to increase the efficient use of available resources (Colla and Rouphael, 2015; Parnell et al., 2016). Maximizing profit by increasing production led to an intensive use of the existing fields (Chellemi et al., 2016) resulting in an increase in infestation and damage by various pests (Gamliel, 2010). Plant disease control is therefore still a "pressing need" for agriculture in this century; controlling diseases that reduce crop yield is required (Emmert and Handelsman, 1999). Chemical pest management in crops is becoming a challenging task, due to the strong requirement to minimize or avoid the use of pesticides (Gamliel, 2010; Singh and Singh, 2009). In addition, resistance to synthetic pesticides is also an increasing problem (Chandler et al., 2011; Droby et al., 2016; Grabke and Stammler, 2015; Lefebvre et al., 2015; Mnif and Ghribi, 2015; Siegwart et al., 2015). Indeed, the awareness regarding environmental degradation and health risk associated with chemical pesticides, increasing production costs and crop losses due to diseases even by increased use of pesticides, opened the door to alternative approaches to pest management (Chellemi et al., 2016; Ji et al., 2006; Siegwart et al., 2015).

In Europe, conventional agriculture based on chemical pesticides is currently in transition to integrated pest management (IPM) due to changes in the European legislation which has as objective human health and environment protection (Lamichhane et al., 2017). The European Directive 2009/128/EC on sustainable use of pesticides was an important step in the reduction of utilization of chemical pesticides. This Directive requires, starting 2014, the implementation of the general principles of IPM in which non chemical methods must be preferred and pesticides should have the smallest possible impact on non target organism and the environment (Colla et al., 2012). The Food and Agriculture Organization of the United Nations (FAO) defines IPM as "the careful consideration of all available pest control techniques and subsequent integration of appropriate measures that discourage the development of pest populations and keep pesticides and other interventions to levels that are economically justified and reduce or minimize risks to human health and the environment. IPM emphasizes the growth of a healthy crop with the least possible disruption to agro-ecosystems and encourages natural pest control mechanisms" (FAO, 2018). In Annex III of Directive 2009/128/EC the general principles are explained (European Commission, 2009a). IPM practice helps to reduce insecticide use, but for the management of soil pathogens, its application is more difficult (Chellemi et al., 2016). Against soil pathogens, chemical fumigation has provided great benefits to agricultural production for many years (Gamliel, 2010), but after the withdrawal of methyl bromide, in countries that used it very intensive, some phytopathological problems became difficult to manage (Colla et al., 2012).

The Directive 2009/128/EC implies the need to apply diverse biological and cultural strategies to reduce synthetic pesticides. Diverse techniques are currently available and the ones generally used in organic farming become of increasing interest also in conventional farming. Good agricultural practices help to prevent or reduce disease symptoms, however the use of fungicides, soil fumigants and disease resistant varieties, hybrids or rootstocks, still constitute the major effective tools to manage plant diseases (Tjamos et al., 2010). In IPM the decision to apply plant protections measures is based on threshold levels of harmful organisms based on monitoring results (European Commission, 2009a). Therefore diverse strategies can be applied to control pests to reduce the use of conventional pesticides, going from cultural and physical controls to host resistance including transgenic plants (Eilenberg et al., 2001). Well known are, for example, crop rotation, the use of resistant or tolerant cultivars, mulching, mix cultures, minimum tillage, biocontrol agents, but also hygienically rules and computer supported models for pest alert, which are very important to reduce or to curtail pest (Colla et al., 2012; Gamliel, 2010; Koike and Gordon, 2015). For soil-borne disease management, where all sources of inoculum among the entire disease cycle have to be considered, measures have to include actions such soil disinfestation and sanitation (Gamliel, 2010).

Biological control or biocontrol agents (BCAs) are in line with the principles listed in Annex III (Matyjaszczyk, 2015) and have the potential to become one main pillar of IPM practice (Lamichhane et al., 2017). BCAs may provide an integration or an alternative to chemical pesticides to manage plant disease (Ji et al., 2006; Lamichhane et al., 2017; Mathivanan et al., 2005; Tjamos et al., 2010). Beneficial micro-organisms received increasing levels of attention since the middle of the 1990ties as biofertilizers, to improve plant growth, and as BCAs, to manage plant disease (Mathivanan et al., 2005) because of consumers' concerns regarding the residues of chemical pesticides (Ji et al., 2006; Liu et al., 2014; Nehra and Choudhary, 2015) and the environmental pollution (Mathivanan et al., 2005; Nehra and Choudhary, 2015). Ease to handle and application with standard machinery are important features for growers (Bashan et al., 2014; Berninger et al., 2018).

Plant growth-promoting rhizobacteria viz. micro-organism that can influence plants and their environment

BCAs are part of the biological control system where living organisms are used to suppress pest population or their damaging impact (Eilenberg et al., 2001). BCA are often found in the well known group of plant growth-promoting rhizobacteria (PGPR). PGPR are soil free-living bacteria that inhabit the rhizosphere (zone around the root), the rhizoplane (root surface) and/or live within the root (endophytes) (Ruzzi and Aroca, 2015). These micro-organisms can benefit from the organic compounds released by the roots as carbon and energy source (Chauhan et al., 2015). Generally PGPR (and root colonizing plant beneficial fungi) exert their mode of action: by changes in plant hormonal balances; through production of volatile organic compounds; by increasing the nutrient availability and favour nutrient uptake by plants; by influencing the abiotic tolerance of plants (Ruzzi and Aroca, 2015); by competitive colonization of plant roots; by stimulation of plant growth and/or reduction of plant disease incidence (Haas and Défago, 2005) and by modulating the root architecture (Vacheron et al., 2016; Verbon and Liberman, 2016). Their mode of action classifies them into three bioproducts: biostimulants, biofertilizer and biocontrol agents. **Biostimulants** are defined as substances or micro-organisms, that enhance nutrition efficiency uptake, tolerance to abiotic stress and/or crop quality traits when applied to plants (Chojnacka, 2015; du Jardin, 2015). **Biofertilizers** increase the availability of nutrients and their utilization by the plants (Chojnacka, 2015; du Jardin, 2015; Nehra and Choudhary, 2015). These micro-organisms can be seen as living fertilizers (Chojnacka, 2015) that act for example through nitrogen fixation or phosphate solubilisation (Haas and Défago, 2005). In this case, the deleterious effects of plant pathogens is reduce or even prevented by the indirect plant growth promotion of PGPR, by producing antagonistic substances or by inducing resistance (Beneduzi et al., 2012).

BCAs are living organisms that protect plants against enemies (du Jardin, 2015) and are comprised in the biopesticide group (Chojnacka, 2015; Haas and Défago, 2005). BCA exert their action directly through: antibiotics; by production of enzymes; by secretion of volatile toxic metabolites; by competition for nutrient resources and essential micronutrients; by interfering with the pathogenesis mechanism; by hyperparasitismus and/or indirectly by inducing resistance (Bardin et al., 2015; Beneduzi et al., 2012; Bhattacharjee and Dey, 2014; Robinson-Boyer et al., 2009; Tjamos et al., 2010). It has been found, that BCA can selective compensate the impact of a pathogen on the plant-associated microbiota. This could be originate by direct impact on the microbiota or by indirect impact on the pathogen (Massart et al., 2015a). The most representative species under the PGPR are *Pseudomonas* and *Bacillus* spp. (Nehra and Choudhary, 2015), which are antagonists of recognized root pathogens.

The bacterial genus Pseudomonas

Members of the genus *Pseudomonas* are rod-shaped motile aerobic Gram-negative bacteria characterized by metabolic versatility (Haas and Défago, 2005; Mnif and Ghribi, 2015), due to the presence of a complex enzymatic system (Mnif and Ghribi, 2015), and a genomic G+C content of 59–68% (Haas and Défago, 2005). Fluorescent pseudomonades, effective rhizosphere bacteria, exert beneficial effect on plant growth promotion in addition to disease control (Commare et al., 2002; Mathivanan et al., 2005). Known to survive in both the rhizosphere and the phyllosphere, they can be used as BCA for the management of foliar infection (Rabindran and Vidhyasekaran, 1996) and as effective strategy against soil-borne disease (Nandakumar et al., 2001). To control soil-borne diseases, pseudomonades produce secondary metabolites and antifungal compounds, compete for nutrients, or produce lytic enzymes that act on fungal cell wall components (Commare et al., 2002). Additionally, they can induce systemic resistance and thus protect the leaves when foliar diseases are controlled by application of the bacteria as seed, soil or root treatments (Vidhyasekaran et al., 1997).

Among fluorescent pseudomonades, *Pseudomonas fluorescens* play an important role in biological control of pathogens. They dominate in the rhizosphere and possess several properties that allow them to be considered the prime candidates for biological control. *P. fluorescens* has excellent root-colonizing abilities, grow rapidly and is able to exploit root exudates, and colonise plant roots of several crops. They can significantly increase yield and enhance plant growth which is often accompanied by reductions of pathogenic fungi and bacteria in the root zone population (Gade and Armarkar, 2011). *P. fluorescens* are interesting bacteria also because of their catabolic versatility, and their capacity to produce diverse antifungal metabolites (Mukherjee and Babu, 2013).

Pseudomonas derived bioactive compounds include: 1. biosurfactants like lipopeptide and rhamnolipid with their efficient antifungal activities; 2. mycolytic enzymes e.g. protease, lipase and glucanase; 3. secondary bioactive metabolites like hydrogen cyanide, salicylic acid and iron chelating siderophores with the ability to reduce plant fungi growth and infection; 4. bacteriocine like the lectin-like bacteriocin produced by *P. fluorescens* Pf-5 and active against a wide variety of phytopathogenic fungi and bacteria. In addition *Pseudomonas* derived biosurfactants and secondary metabolites like phenazines and siderophores can induce systemic resistance in plants (Mnif and Ghribi, 2015). It is also reported that *Pseudomonas* biocontrol strains were observed at the root surface, forming micro-colonies or discontinued biofilms, capable of endophytic behaviour in the intercellular spaces of the epidermis and the cortex (Maurya et al., 2014).

Diverse *P. fluorescens* products were developed and tested in recent years. Two *Pseudomonas* strains are listed in the European Union (EU) as active substances: *P. chlororaphis* MA 342 and *Pseudomonas* spp. strain DSMZ 13134. *P. chlororaphis* MA342 is authorized at a national level in 14 nations, DSMZ13134 in 15 nations and in one the authorization is in progress (EU, 2018). In Germany, both are approved as fungicide (Table 2). *P. chlororaphis* MA 342 is sold in two products, as emulsion or flowable concentrate for seed treatment, for spelt, barley, rye, triticale and wheat. It is used against *Tilletia caries*, *Tilletia foetida*, *Fusarium* spp., *Pyrenophora graminea*, *Pyrenophora teres* and *Septoria nodorum*. *Pseudomonas* spp. strain DSMZ 13134 is sold as wettable powder for potatoes against *Rhizoctonia solani*.

Production and formulation of bioproducts

The production and formulation of bioproducts is a challenge. Many potential highly useful strains are described in the scientific literature, but just a few arrive on the commercial market (Bashan et al., 2014). In 2005, Haas and Défago stated that to find an effective biocontrol PGPR strain is time consuming because no *in vitro* diagnostic kits were available. Nowadays, through the easier access to molecular technologies, screening can be almost made *in vitro*. Beside the classical methods in which the production of secondary metabolisms are tested, polymerase chain reaction (PCR) targeting relevant genes involved in defence traits can be used (Hol et al., 2013; Vacheron et al., 2016). The discovery of the adequate micro-organism is just the first step in the BCA research (Bashan et al., 2014). The following steps to produce a micro-organism ready for commercial use, involve collaboration of diverse disciplines (see figure 1) and generate high costs (Droby et al., 2016; Köhl et al., 2011). Different issues have to be considered for a commercial product: it has to be compatible with field routine practice, storable, adaptable to different

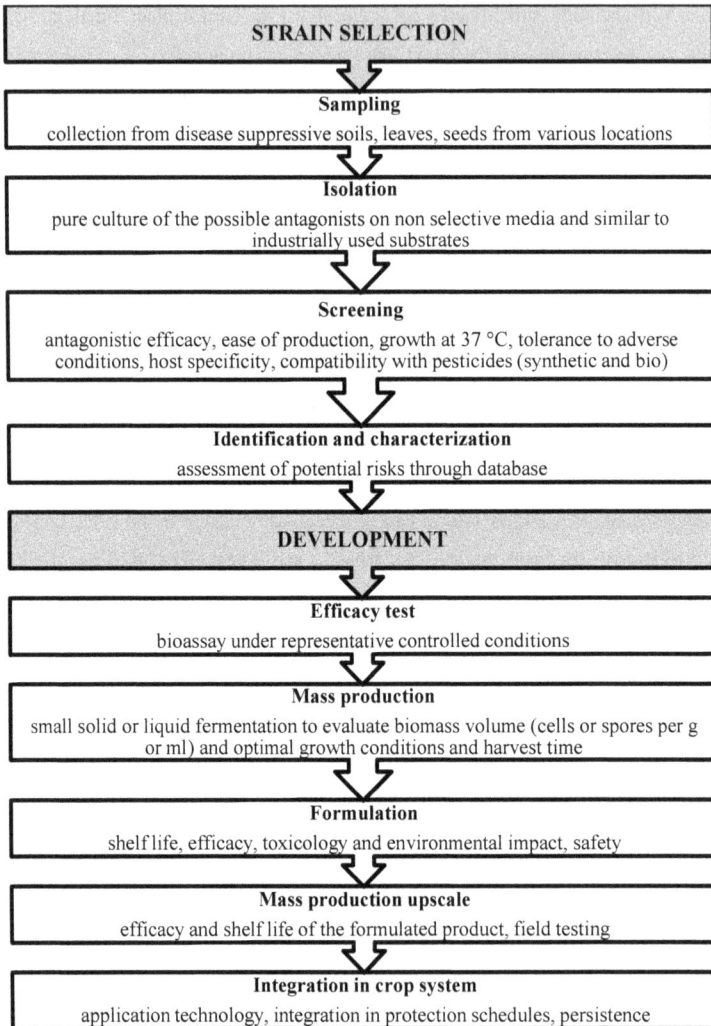

Figure 1: Steps from screening to development of a microbial pesticide for a target crop and disease. For the industry the costs for development are also important but not considered here. Costs depend on the market size, presence of competitive products, availability of the selected micro-organism (patents), regulations and registration costs (Jackson, 1997; Jensen et al., 2016; Köhl et al., 2011; Montesinos, 2003).

field conditions and soil types, it has to provide reproducible results and be safe for humans, animals and plants (Bashan et al., 2014). This means that after finding the best strain for the intended effect on a target crop, its production follows, with regards to the design of a specific formulation (Bashan et al., 2014).

Production of micro-organisms in general, is made by liquid or solid **fermentation**, with some exceptions, for example viruses. Fermentation parameters can influence the efficacy and shelf life of BCA as the following formulation. Solid-state fermentation is a process in which fungi and bacteria are grown on solid material without the presence of free water. Usually, natural solid wastes or by-products containing polysaccharides in their structure are used, making it an efficient and environmentally safe method for micro-organisms mass production. The resulting material is then used for seed coating, encapsulation or direct introduction in soil. It was found that spores of micro-organism produced by solid state fermentation, were more effective in reducing disease symptoms, adhere better on roots and had higher survival than spores produce by submerged cultivation. The solid on which the microbials are grown, can act as carrier for cells or spores, reducing the following formulation process. In addition, it is possible to grow more micro-organisms together with the purpose of a dual or more inoculation (Vassilev et al., 2015). Nevertheless, in the industry, submerged or liquid fermentation is the most widely used system and ideal for bacteria (Ash, 2010). It is easy to scale up and through centrifugation or filtration it is possible to concentrate cells and spores (Jones and Burges, 1998; Montesinos, 2003). In addition, submerged fermentation is used as a cost-effective and easy technology for the mass production of secondary metabolites, antibiotics and/or organic acids (Keswani et al., 2016).

To **stabilize the viability** of the obtained product, refrigeration, or dehydration, are the preferred methods. Dehydration allows optimum conditions for storage, handling and distribution of microbial products. In fact, since decades this is the preferred preservation method but there is no generic drying system for all applications (Morgan et al., 2006). At industrial level, where low production costs are also important, the preferred methods are spray- or fluidized bed-drying (Montesinos, 2003). Spray-drying uses high temperatures to dry large quantities of cultures in a short time (Droby et al., 2016; Palazzini et al., 2016). The costs are low, however this method is only adapt for BCA that produce endospores that are heat resistant. For heat-sensitive micro-organisms, fluidized bed-drying could be used because the drying temperatures are relatively low (Droby et al., 2016). However for micro-organism, as for example *Pseudomonas*, high loss in viability, can still be a consequence due to the thermal treatment (Montesinos, 2003). Freeze-drying maintains high cell viability for bacteria, yeasts and sporulating fungi but is more costly than the other drying processes (Droby et al., 2016; Montesinos, 2003). It assures long viability, storability, protection from contamination during storage and ease of product distribution (Miyamoto-Shinohara et al., 2000; Smith and Onions, 1983). Freeze-drying is commonly used to preserve bacterial cultures in research and industry (Morgan et al., 2006; Palmfeldt et al., 2003). The tolerance of micro-organisms to survive freeze-drying vary greatly: Gram-negative bacteria show lower survival rate than Gram-positive bacteria and spores are more resistant than vegetative cells (Morgan and Vesey, 2009). Beside the bacterial species, production condition and the freeze-drying parameters influence bacterial survival during the process (Heckly, 1985). In fact structural and physiological injury of the bacterial cell can occur during freezing, freeze- and air-drying, which can result in considerable viability loss (Prasad et al., 2003). Increase in viability can be achieved by the addition of substances that protected cells during the process like cryo- (CPAs) or lypo-protectants (LPAs) (Morgan and Vesey, 2009). Added before freeze-drying, these protective agents have an important role in the protection of the bacterial cell membrane and cytoplasm against dehydration during the process (Hernández et al., 2006).

The **type of application form** influences its use. For example, diverse talc-based formulations of *P. fluorescens* were developed, however these are not adaptable when micro irrigation is used for plant diseases management (Manikandan et al., 2010). In this case, a liquid formulation has several additional advantages including high cell count, zero contamination, longer shelf life and greater protection against environmental stresses (Manikandan et al., 2010). These formulations are a simple way to stabilize the microbial viability (Droby et al., 2016). Liquid formulations are flowable or aqueous suspensions containing micro-organisms in water, oils, or their combinations (Berninger et al., 2018; Schisler et al., 2004), with the addition of substances to improve stabilization or dispersal abilities. Nutrients, cell protectants and other substance that can improve performance can also be added. Dry formulation can have diverse forms depending on the size of the end product and with which carrier the active ingredient is mixed (Bashan et al., 2014; Droby et al., 2016). Granules, for example, are made by a carrier coated with the active ingredient, vice versa for a capsule, where the active ingredient is coated with the carrier. Wettable powders are easy to produce and water dispersable granules mix fast with water and produce less fine dust because a binder holds all together (Jones and Burges, 1998; Schisler et al., 2004). In dry formulations, the micro-organism remains in a dormant form:

the metabolism is slowed down or even stopped so they are resistant to environmental stresses and insensitive to contamination (Bashan et al., 2014). Solid or liquid formulated microbials can be applied on seeds by mixing them by hand or with cheap to operate machinery like cement mixers. After drying the seeds can be sown (Bashan et al., 2014). When used for soil inoculation, this could be done with machinery during planting or sowing for solid formulation, by root dipping for liquid formulation or by irrigation, also during growing season (Berninger et al., 2018).

In addition to an easy handling qualities and a reasonable price, repeated positive results and long storage quality are important. Formulation influences the biopesticides **performance**. Indeed, a major constraint is, that BCAs, but also other bioproducts like biofertilizer, are often less consistent in their performance than chemicals (Massart et al., 2015a). Often this inconsistency is caused by inappropriate formulations (Paau, 1988), because downstream processing may have a direct or indirect detrimental effect on the properties of the selected BCA (Berninger et al., 2018; Droby et al., 2016). Therefore, during BCA selection, the efficacy should be tested with the end product to avoid bad performance when the product comes to the market. To improve field efficacy and shelf life, the addition of various compounds, to the active agents, was a solution for the formulation technologies (Droby et al., 2016; Siegwart et al., 2015). For liquid formulations conflicting information about **shelf life** are reported with longer or shorter life as solid carrier-based products, depending on the case (Bashan et al., 2014). Liquid formulation require refrigeration to stabilize cell viability while dry formulations can be stored at room temperature (Droby et al., 2016).

As seen, many steps follow the discovery of the best microbial antagonist and each influences the end performance of the BCA in a product. Indeed, efficacy depends strongly on the formulation, the formulation depends on the fermentation type (Abadias et al., 2001; Spadaro and Gullino, 2005) and culture conditions can influence the end characteristics of microbial strains (Fuchs et al., 2000; Wessendorf and Lingens, 1989). In biopesticide production, the fermentation of the active ingredient and the formulation used are often closely linked (Hynes and Boyetchko, 2006). To understand how a BCA behaves helps in the selection of the best strain, and can help to improve its efficacy and consistency by designing the most appropriate formulation. In addition, these knowledge can assist the grower in optimizing the method and timing of application (Massart et al., 2015b).

Biopesticides, plant protection products and the new regulation

The use of biopesticides to reduce the amount of synthetic chemical products increased in recent years, though they have not yet become major players on the pesticide market (Gamliel, 2010). One reason is that their price is still high, because at present the use of biopesticides is mainly done in minor agricultural and horticultural situations (Gamliel, 2010; Siegwart et al., 2015). Anyway, their market share increased about 10% between 2005 and 2010, while a decrease of 1.5% was registered for synthetic pesticides (Siegwart et al., 2015). Biopesticides are often considered to be: not reliable, to lack in robustness and to require a high technical level for their utilization, because they are generally more sensitive to environmental variations than synthetic pesticides (Siegwart et al., 2015). Regardless of these issues, biopesticides undergo the same government regulations for registration and use as synthetic chemical plant protection products (PPPs). Biopesticides are considered PPPs in most countries due to the Regulation (EC) No 1107/2009 of the European Parliament and of the Council concerning the placing of plant protection products on the market and repealing Council Directives 79/117/EEC and 91/414/EEC (Köhl et al., 2011; Lamichhane et al., 2017; Matyjaszczyk, 2015). The regulation refers to PPP as the end product (containing active substances, safeners or synergists) applied to plants or plant products to: protect against all harmful organisms; influence the life processes of plants; preserve plant products and destroy or prevent growth of undesired plants or parts of plants. The term "active substance" refers to substances that have a general or specific action against harmful organisms or on plants including their parts products (European Commission, 2009b). According to that definition, fungi, bacteria, algae, protozoa viruses, pheromones and semiochemicals, and plant extracts/botanicals included in biopesticides (EU, 2018; OECD, 2018) are considered active substances. As a consequence BCA, which includes micro-organism defined as microbiological units able to replicate or to transfer genetic material (European Commission, 2009b) are part of the biopesticide. The Regulation (EC) No 1107/2009 is legally binding and the registration of PPPs is carried out according to its rules (Matyjaszczyk, 2015). Within these rules the registration procedure has two levels: 1. the active substance is autorized and listed in the register of active substances as safe enough to be used in plant protection after evaluation on the basis of experiment results; 2. PPPs that contain the approved active substances, are place on the market of member states if accepted as sufficiently safe and efficient after evaluation on the basis of experiment results. A difference in the registration requirements between chemical and microbial pesticides is, that for micro-organism the achievement of the data collection, regarding fate and behaviour in the environment, can be based on the open literature. This is also applicable to the data regarding residue exposure (Matyjaszczyk, 2015). Also the description of the mode of action is required for the registration under EU regulations (Massart et al., 2015b). Antagonists that produce secondary metabolites to inhibit the targeted pathogens in *in vitro* assays, are generally excluded, based on the assumption, that indications of antibiotic production would be problematic in the registration process (Droby et al., 2016). If BCAs could be placed within the low-risk category, their registration would be easier and less expensive (Lamichhane et al., 2017). In 2016 a draft to implement the regulation amending the criteria of low-risk substances was initiated. There, micro-organism may be considered low-risk active substances unless multiple resistance to anti-microbials used in medicine is shown (EU, 2018).

Microbial products that have no direct action against pests are placed on the market as plant biostimulants (Colla and Rouphael, 2015), plant strengtheners, plant growth stimulants or plant growth improvers (Matyjaszczyk, 2015). Biostimulants definition is not accepted by all regulatory bodies and these are generally placed on the market as fertilizers according to the rules set by member states (Colla and Rouphael, 2015; Matyjaszczyk, 2015).

Microbial biopesticides present on the German market

The biopesticide market increased significantly worldwide from 0.2% in 2000 to 4.5% in 2010 (Keswani et al., 2016). A market increase of 12.3% is forecasted between 2012 and 2020, versus the 5% estimation for chemical pesticides (Droby et al., 2016). According to the Register of Plant Protection Products of the Federal Office of Consumer Protection and Food Safety (BVL), the official source of information for the German public, out of 1596 products, 272 active substances have been authorized in plant protection in Germany up to now. However, a limited number of micro-organisms are included, about 9% of the total active ingredients (an increase of about 2% since November 2017). In particular, in fruit growing, the authorized active substances, present in 526 PPPs, are 112. From that, 25 micro-organisms are used to produce 38 products. For organic strawberries for example, 21 products are allowed (5 bactericide, 5 fungicide and 11 insecticides) during growing season, however the active substances are just 9. For conventional strawberries the number

Table 1: Overview of the total active substances authorized in PPPs in Germany with particular attention to strawberries.

Function	PPP	AS	PPPs containing MO	MO	AS EU*	AS Germany*
PSM	1596		38	25	493	272
Fruit growing	526		13	11		
	Strawberries				**All crops**	
Acaricide	8	5			39	13
Acaricide+Insecticide	18	3	1	1		
Bactericide	5	2			13	7
Fungicide	70	26	3	4	156	94
Herbicide	24	14			123	89
Insecticide	29	10	1	1	108	50
Molluscicide	70	2			2	2
Repellent	5	1			17	7
Rodenticide	22	2			7	5
Elicitor					9	1
Plant activator					2	
Plant growth regulator	1	1			44	25
Soil treatment					1	

PSM: all registered PPP; AS active substance; MO micro-organism
BVL online (21.03.2018) with the selection strawberry and fruit growing
https://apps2.bvl.bund.de/psm/jsp/index.jsp
* Active substances approved in PPPs for all crops: EU pesticide database (21.03.2018)
https://ec.europa.eu/food/plant/pesticides_en

of pesticides used is 247, made out of 64 active substances (table 1). Out of these, three are made of micro-organisms (*Ampelomyces quisqualis* strain AQ10, *Bacillus subtilis* QST 713 and *Metarhizium anisopliae* var. *anisopliae* F52) which are also accepted in organic strawberry growing. The general acceptance of BCAs is still problematic. In an interview performed in 2008, Moser et al. reported the results of a survey of North Rhine-Westphalia farmers, representing 30% of the strawberry production area of the region, about their preference for BCAs: only 70% of the interviewed farmers trust them, however almost 20% of the skeptics used them anyway (Moser et al., 2008).

In table 2 the micro-organisms that are accredited in Germany as active substance for PPPs are listed. From this list, diverse active substances were used in *in vitro* screening to select "the best one's" for diverse trials that are presented in this thesis. The entomopathogenic fungus **Metarhizium brunneum** is a fungus with a wide host range however confined to insects and some Acarina species. In particular *M. brunneum* Ma43 is found in the EU pesticides database under active organism *M. anisopliae* var. *anisopliae* BIPESCO 5/F52. It was isolated in Austria from *Cydia pomonella* (codling moth) and its use is authorized in Germany since August 2014 against weevils of the genus *Otiorhynchus* (European Commission, 2014). Beside its activity against insects, this strain

Table 2: Micro-organisms listed as active substances in PPPs in Germany.

Active substance	Product name	Type	Fields of use	Expiration date authorization
Adoxophyes orana granulovirus BV-0001	CAPEX 2	I	Fruit growing	31.12.2022
Ampelomyces quisqualis strain AQ 10	AQ 10 WG	F	Vegetable and fruit growing	31.07.2018
Aureobasidium pullulans DSM 14940 and *Aureobasidium pullulans* DSM 14941	Blossom Protect // Botector	B, F	Fruit growing // Viticulture	31.12.2025
Bacillus amyloliquefaciens QST 713 (formerly *B. subtilis*)	Serenade ASO	F	Vegetable and fruit growing	30.04.2019
	Serenade MAX	B	Fruit growing	31.12.2020
Bacillus thuringiensis spp. *kurstaki* ABTS-351 (strain HD-1)	Dipel ES // Universal-Raupenfrei Lizetan // Lizetan Buchsbaumzünslerfrei // BACTOSPEINE ES	I	Field crops, forestry, vegetable, fruit and ornamental growing, viticulture	31.12.2021
Bacillus thuringiensis spp. *kurstaki* strain EG-2348	Lepinox Plus	I	Vegetable and fruit growing, viticulture	30.04.2020
Bacillus thuringiensis ssp. *aizawai* ABTS-1857	FLORBAC // Lizetan Raupen- & Zünslerfrei // XenTari // Xentari BuchsbaumzünslerFrei // Xentari RaupenFrei // Zünsler & Raupenfrei Xentari	I	Vegetable, fruit and ornamental growing, viticulture	30.04.2020
Bacillus thuringiensis ssp. *aizawai* GC-91	Turex	I	Vegetable and ornamental growing	30.04.2020
Bacillus thuringiensis ssp. *israelensis* (serotype H-14) AM65-52	Gnatrol SC // Neudomück Pro	I	Ornamental growing	30.04.2019
Bacillus thuringiensis ssp. *tenebrionis* NB 176 (TM14-1)	Novodor FC	I	Field crops	31.12.2022
Beauveria bassiana ATCC 74040	Naturalis	I, A	Vegetable growing	31.12.2024

25

Table 2: continued.

Active substance	Product name	Type	Fields of use	Expiration date authorization
Coniothyrium minitans CON/M/91-08	Contans WG	F	Field crops, vegetable and ornamental growing	31.12.2018
Cydia pomonella granulovirus isolate GV-0006	Madex MAX	I	Fruit growing	31.12.2021
Cydia pomonella granulovirus isolate GV-R5	CARPOVIRUSINE EVO 2	I	Fruit growing	30.04.2020
Cydia pomonella granulovirus mexican isolate	CARPOVIRUSINE	I	Fruit growing	31.12.2022
Gliocladium catenulatum J1446	Prestop // Prestop Mix	F	Vegetable and ornamental growing	31.07.2018
Metarhizium anisopliae var. *anisopliae* F52	Met52 Granulat Met52OD // Bio1020OD	I A, I	Fruit, vegetable and ornamental growing	30.04.2019
Pepino mosaic virus strainCH2 (isolate 1906)	PMV-01	V	Vegetable growing	05.12.2030
Pseudomonas chlororaphis MA 342	Cedomon // Cerall	F	Field crops	30.04.2019
Pseudomonas spp. strain DSMZ 1314	Proradix	F	Field crops	31.01.2025
Pythium oligandrum M1	Polyversum	F	Field crops	30.04.2020
Trichoderma asperellum ICC 012 (formerly *T. harzianum*) and *Trichoderma gamsii* ICC 080 (formerly *T. viride*)	Bioten	F	Vegetable and ornamental growing	31.12.2024
Verticillium albo-atrum isolate WCS85	Dutch Trig F	F	Ornamental growing	30.04.2020

I insecticide; F fungicide, B bactericide; V viricide; A acaricide
BVL online (21.03.2018) https://apps2.bvl.bund.de/psm/jsp/index.jsp

26

increased the yield of grade one strawberries significantly and reduced the quantity of deficient fruits during field trials (Bisutti et al., 2013). *Trichoderma* is another well known fungus also present in table 2 as active substance. *Trichoderma* spp. are free-living fungi, characterized by rapid growth and high interaction capability in root, soil and foliar environments. These fungi are efficient mycoparasites and competitors, who produce antibiotics as well as a wide range of enzymes (Harman, 2006) and a large variety of volatile secondary metabolites (Singh and Singh, 2009). These characteristics make *Trichoderma* spp. an interesting BCA, a systemic resistance inducer in plants and a plant growth promoter also through increased nutrient uptake (Akhtar and Azam, 2014; Harman, 2006). *Bacillus*, a spore-forming Gram-positive bacteria, is the most represented bacteria in table 2. In addition to their insecticidal and nematocidal activity, *Bacillus* spp. exhibit also an antimicrobial activity through secondary metabolites, mycolitic enzymes and biosurfactants. For example, *B. amyloliquefaciens* FZB42, a good root colonizer, produces secondary metabolites with antimicrobial activity and was able to reduce the effect of pathogen *R. solani* on lettuce seedling by direct interaction or systemic resistance effects. However, this bacterium is better known for its beneficial effects on plant growth (Chowdhury et al., 2013).

The plant and its rhizosphere microbiome

Plants harbour a very diverse and plentiful micro-organism community (microbiota) on roots and leaves. These communities provide specific functions and features to the plant, influencing health and productivity by producing compounds that affect plant gene expression, root architecture and plant defence response. On the other hand, the plant selects its own microbiota by producing exudates, on which micro-organisms react and regulate gene expression, and through leaf and root constitution (Hartmann et al., 2009; Massart et al., 2015a; Venturi and Keel, 2016). Micro-organisms communicate together by quorum sensing (Ambrosini et al., 2016; Trabelsi and Mhamdi, 2013; Venturi and Keel, 2016) so that they can form and synchronize their behaviour (Ambrosini et al., 2016; Venturi and Keel, 2016). Plant and soil microbial communities differ significantly in species composition, abundance and diversity. The rhizosphere microbiome is mainly influenced by soil type (composition and physical and chemical properties) and plant genotype (by root morphology, physiology and exudates composition, and the presence of defence genes). Plant species (morphology and chemical composition) and genotype (mutations in a plant gene) influence the phyllosphere microbial community. Abiotic factors like environmental origin, growing season, fertilizers, crop rotation and pesticides can influence these microbial communities too (de Souza et al., 2015; Massart et al., 2015a). Plants are able to distinguish between beneficial microbiota and pathogen by recognizing molecules released by micro-organism through different receptors in the plant cells (de Souza et al., 2016). Interactions between plants and bacteria can be symbiotic, endophytic or associative depending on the vicinity to roots or rhizosphere (de Souza et al., 2015). The form of relationship is established due to the numerous molecular signalling events between them (Ambrosini et al., 2016). Endophytes are micro-organisms that colonize internal plant tissue and cause no negative effects on plant growth (de Souza et al., 2015). Symbionts produce molecules that trigger signals cascades that lead to infection and accommodation of the symbiont. Symbionts and pathogens compete frequently on plant tissue and in the rhizosphere by producing antimicrobial molecules. Produced by the pathogen, these secondary metabolites can even change the structure and abundance of soil microbial populations with problematic consequences for plant and micro-organism (de Souza et al., 2016; de Souza et al., 2015). Inoculating the soil with BCA can change the microbial community, although modification can be buffered by ecosystem resilience. Even in case some bacterial species are lost, ecosystem functions are maintained through the high content of bacteria where one function may be carried by different ones (Ambrosini et al., 2016; Trabelsi and Mhamdi, 2013; Vassilev et al., 2015). A strain may act with several modes of action but, it is not yet clear which conditions influence the switch of action mode and if there is one dominant (Bardin et al., 2015). Anyhow, inoculation with two or more strains can contribute to better performance and soil health through cooperation between organisms (Ambrosini et al., 2016). However, co-inoculation does not necessary bring an additive or synergic effect. Through competition, the expected positive influence obtained by single strain inoculation can be reduce or even disappear (Trabelsi and Mhamdi, 2013).

Plants are influenced by abiotic and biotic factors. The interaction between plants, macro and micro-organisms and environment is a complex field. To understand function and nutrients cycle, in different ecosystems like agricultural ones, plant/micro-organism interactions are of enormous importance (Berg and Smalla, 2009). Combination of traditional techniques and novel cultivation-independent methods can be used, depending on the information wanted (Berg and Smalla, 2009). In recent years, developments in high-throughput sequencing technologies and in bioinformatics helped to increase the understanding about soil microbial communities (Droby et al., 2016; Massart et al., 2015a) and can also help to understand the influence that the addition of BCA has on plant-microbiome system. The analysis approach can be addressed to a single target or be more holistic. Information can go from taxonomic description, to gene composition and expression and functional understanding (Massart et al., 2015a). With denaturing gradient gel electrophoresis it is possible to follow changes of microbial communities in time and space, however samples with high degree of diversity in the community composition could be problematic. Real time PCR and terminal restriction fragment length polymorphism can be used to assess total microbial communities or their diversity and structure (Trabelsi and Mhamdi, 2013). Information about the microbial community genomes in their composition and expression with a more holistic approach can be done by next generation sequencing (NGS). Amplicon, the most popular, and metagenome sequencing give information about the microbial community and about the gene content of a population respectively. These can help to describe more accurately and improve understanding of the role of the microbiome in biocontrol of plant pathogens. A functional supplement to metagenome is the metatranscriptomic approach that identifies the most transcribed genes and pathways (Massart et al., 2015a). Proteomics can be used to identify the biocontrol properties of a potential BCA. Indeed, basic biological responses through the qualitative and quantitative analysis of the expressed proteins can be understand. Proteins with biotechnological value or differentially expressed in relation to the biocontrol process can be identified (Massart et al., 2015b). NGS will increase knowledge about microbiome development and overtime evolution (Massart et al., 2015a). Additionally it can be used to assess physiological changes and the effect of environmental stress on BCAs intracellular machinery (Droby et al., 2016). In fact, suppression showed *in vitro* is not direct reportable to complex population systems as shown for pseudomonads populations hosted in the rhizosphere in their ability in plant pathogen suppression (Gardener, 2007).

Here, macro-biota, farming management and environment were not considered, however also these have a huge impact on the plant and its microbiome. Farming management can alter the soil environment (Ambrosini et al., 2016; Gardener, 2007) which can result in intensive soil degradation and progressive loss in fertility. The microbial communities of agricultural ecosystems are constantly disturbed by the management techniques which affect their function (Ambrosini et al., 2016) and dynamics (de Souza et al., 2015) as shown for biocontrol pseudomonads in their relative abundance (Gardener, 2007). As seen, microbial community composition and dynamics depend on multiple factors like plant species and physiological state and/or environmental conditions

(Ambrosini et al., 2016; Massart et al., 2015a). These factors will also take directly influence on the behaviour of an added BCA, in addition to the influence that microbiota has on BCA survival and activity, and indirectly by stimulating plant defence (Massart et al., 2015a).

Strawberry

The modern strawberry (*Fragaria* x *ananassa* Duchesne) is a result of hybridization between the two wild species *F. virginiana* and *F. chiloensis,* started around 1750. It is now the predominant cultivated strawberry. *Fragaria* spp. is a herbaceous, perennial member of the Rosaceae, adapted to widely varying environmental conditions and resistant to several diseases and pests. The crown is a shortened stem in which leaves and axillary buds are inserted in a restricted area. The buds can develop to flowers, runners or branch crown depending on environmental conditions (Hancock, 1999; Maas, 1998). Shorter day light and lower temperature normally induce flower bud development, whereas longer days favour vegetative growth and stolon production (Guttridge, 1959). The roots are generally 30 cm long, a mix between perennial structures and feeder roots, short-living or annual roots, responsible for water and nutrient uptake (Hancock, 1999; Maas, 1998). Strawberries can be produced on different types of soil that range from sand to heavy loams. The best soil pH is between 6.0 and 6.5 although they are tolerant to a wide pH range. To achieve high yields plants should grow in deep fertile soil, with high organic matter and good drainage (Hancock, 1999).

Strawberry market and organic farming

Strawberries are a beloved fruit and Germany was until 2015, under the 10 top producers worldwide. In 2016, about 8.6% of the European production was harvested in Germany. The worldwide production was more than 9.1 million tons and of these, Europe produced 18.3% (FAO, 2018). In Germany in 2017, from the 17,807 hectare of strawberries distributed on 2,250 farms, 294 hectares were organic. Out of 135,283 tons of total strawberries, 1,856 tons were organically grown. The German strawberry organic production decreased in the last years (Destatis, 2018), but the requirements of the consumer for organic products is increasing. For fresh strawberries, already in 2006, Tahmatsidou et al. reported that organic production is an expanding market, but due to the reduced yield and higher price the authors conclude, that it will not be practical for the mass market and processing sector (Tahmatsidou et al., 2006). Anyhow, consumers spend more money to buy organic, because organic growing permits to produce sustainable, with respect to plants and animals, and delivers food with less to no pesticide residues (Reganold and Wachter, 2016). Germany is the main consumer of organic products in the EU, only behind USA in the world list. 5.0% of the food market business volume was spent for organically grown products. Experts assess a clear potential increase for this market. In Germany the number and area of organic farming has increased constantly since 1996. In 2016 7.5% (1,251,320 hectare) of the whole crop area and 9.9% (27,132) of the whole farm number produced organically (Bmel, 2018). Organic farming is a system in which food production has the minimal possible impact to ecosystems and is often proposed as a solution when a reduction of environmental harm is in focus (Seufert et al., 2012). In this system, the use of synthetic fertilizers and pesticides are avoided although yield is reported to be lower than in conventional cropping systems (Conti et al., 2014; Esitken et al., 2010; Martin and Bull, 2002). For organic systems the yield reduction is 25% in contrast to conventional systems with differences across crop types and species: organic fruits show 3% reduction, vegetables 33% (Seufert et al., 2012). Anyway, the substitution of chemical with biological control can be highly favourable when looking to the cost-benefit ratio, depending on the crop (Lefebvre et al., 2015; Reganold and Wachter, 2016). Indeed, it has been shown, that integrated farming systems, that blend some conventional with mostly organic practices, are more sustainable than conventional ones (Reganold and Wachter, 2016).

Overview of the major strawberry diseases

Strawberry growth and fruit production can be influenced by abiotic and biotic factors. For example strong cold periods can injure the rhizome (figure 2a) or the flower (figure 2b) and an inappropriate resource management can favour pests and diseases. Macro pests that can affect strawberries are for example aphids, strawberry blossom weevils (*Anthonomus rubi*), dwelling nematodes, as *Pratylenchus penetrans,* or mites like *Phytonemus pallidus*. In addition, strawberry plants and fruits are sensitive to various diseases caused by fungi and bacteria, but also viruses and phytoplasma. All these pests can reduce yield directly by infecting the fruit or indirectly by weakening the plant, causing it to die.

Botrytis cinerea is the causal agent of grey mould, one of the major fungal diseases of strawberries worldwide. It is an ubiquitous necrotrophic pathogen that can infect flowers, fruits and leaves. Diseased flowers may not produce fruit or the infection can remain quiescent until fruit ripening. Infected fruits and leaves are often covered with grey mould (mycelia, conidia etc.) and fruit can rot (Elad et al., 1996; Freeman et al., 2004). Debris left in the field, like mummified fruit and leaves serve as the main inoculum source (Freeman et al., 2004). The infection is supported by high humidity, free moisture on surface and temperatures between 10 and 20 °C (Elad et al., 1996; Grabke and Stammler, 2015). *B. cinerea* conidia can also germinate at lower temperatures but need longer period for infection (Helbig, 2002). Therefore, grey mould can be a problem also for cold stored picked fruits (figure 2c). *B. cinerea* is known to have high genetic diversity and can easily adapt to chemical control strategies, however isolates with more than one resistance mechanisms have lower fitness (Grabke and Stammler, 2015). It was also found that several *B. cinerea* strains have a significantly reduced sensitivity to the antimicrobial compound pyrrolnitrin produced by several BCAs (Bardin et al., 2015). This shows, that it is necessary to use as many modes of action as possible to control grey mould (Grabke and Stammler, 2015).

Podosphaera aphanis causes powdery mildew, a serious disease especially in tunnel protected production and in warm and dry climates (Maas, 1998). All above ground parts of the strawberry plant, especially leaves, can be infected and this is shown with whitish and powdery spots (figure 2d and e). The disease can decrease fruit set, fruit can show cracks and deformations and poor flavour development. Storage of fruit can be influenced negatively (Pertot et al., 2008). Reduced light intensity, high relative humidity without free moisture and temperatures between 15 and 27 °C favour the disease development (Maas, 1998).

The pathogen *Phytophthora cactorum* is responsible for leather and crown and root rot. It is an Oomycetes that survives as oospores which in presence of free water liberate zoospores produced in sporangia (figure 2h and i). Zoospores can infect immature and mature fruit when splashed on the fruit, or enter wounds and infect plants (Maas, 1998). The spores persist in the soil for a longer time or in infected plants (Eikemo et al., 2003). The young leaves wilt often suddenly and subsequently, the rest of the plant collapses and dies, typically within a few days (figure 2f). By dissecting the crown, brown discoloration can be revealed, a characteristic of the disease (figure 2g). To cause infection, the weather has to be warm and wet. The disease is enhanced by high temperatures and when the plant is water stressed (Maas, 1998).

Verticillium wilt, another soil-borne disease, is caused by *Verticillium dahliae* and *V. albo-atrum* (Maas, 1998). *V. dahliae* has a wide host range (Klosterman et al., 2009; Yang et al., 2014), occurs throughout the temperate zones of the world and the wilt is favoured by environmental stresses (Maas, 1998). The infection affects plant growth (Mol et al., 1996). The outer strawberry leaves can collapse with green and turgid, but stunted inner leaves until the plant dies (Maas, 1998) (figure 2j). *V. dahliae* produces

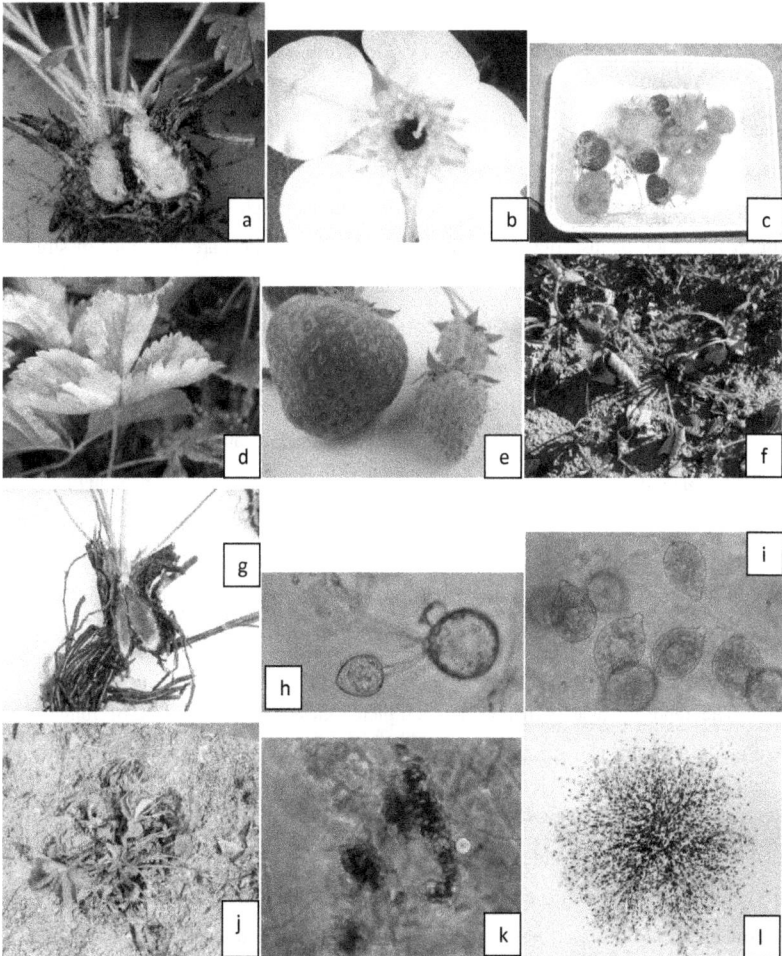

Figure 2: Different damages on strawberry plants: freeze damage on rhizome (a) and flower (b); grey mould (*Botrytis cinerea*) on stored fruits (c); powdery mildew (*Phodosphera aphanis*) on leaves (d) and fruits (e); crown rot symptoms on plants (collapsed plant) (f) and browning of the rhizome (g); germinating oospore producing sporangia (h) and papillate sporangia (i) of *Phytophthora cactorum*; Verticillium wilt symptoms on plant (j); *Verticillium dahliae* MS (k) where each individual cell can germinate alone; germinated MS on pectate medium agar with new MS in colony form (l).

microsclerotia (MS) (figure 2k and l) as resting structures that can survive up to 15 years in the soil (Klosterman et al., 2009). The soil MS density depends mainly on the cropping history, since infested host debris (Mol et al., 1996) remain in the soil after the plants degrade and can be disseminated by wind and water (Maas, 1998). Especially in hot climates, wilt caused by *Verticillium* is most severe in irrigated fields (Uppal et al., 2008).

Another important wilt is caused by ***Fusarium oxysporum* f. sp. *fragariae*.** The disease occurs mainly on well-established plants with wilt of the older leaves that change colour to grey green and begin to dry. With exception of the central young leaves all the foliage collapses when the disease progresses and the plant can die. Diseased crowns show dark brown to orange brown vascular on cortical tissue. High temperatures, water stress through saturation or insufficient irrigation, poor soil conditions are examples of stresses that can cause a more rapid and severe disease development (Koike and Gordon, 2015).

Other known diseases are ***Colleotrichum acutatum*** that causes anthracnose and can attack all parts of the plant, causing serious effects (Maas, 1998); ***Gnomonia fragariae*** causes severe root rot and petiole blight of strawberry (Moročko-Bičevska and Fatehi, 2011); ***Diplocarpon earliana*** and ***Mycosphaerella fragariae*** cause strawberry leaf scorch and spot respectively. The bacterium ***Xanthomonas fragariae,*** that causes angular leaf spot, and the Oomycetes ***Phytophthora. fragariae* var. fragariae,** responsible for red core or red stele, are in the "A2 list of EPPO that recommends to regulate them as quarantine pests" (EPPO, 2018). In recent years ***Macrophomina phaseolina*** causal agent of charcoal rot and ***Fusarium solani*** causal agent of crown and root rot, have emerged (Pastrana et al., 2016).

The reported disease list is not complete; these are only a part of fungal and bacterial diseases that can attack strawberries.

Management of strawberry soil pathogens

Soil pathogens are responsible for dramatic yield losses (Berg, 2006; Martin and Bull, 2002) and increased soil infestation. An accurate determination of the disease is important to allow application of the right pest management. For some diseases a field diagnosis is difficult and a laboratory test would be necessary to identify the causal agent (Koike and Gordon, 2015). One example is the collapsing of strawberry plants that can be caused by four soil pathogens that have similar if not identical symptoms. *Fusarium*, *M. phaseolina*, *V. dahliae* and *P. cactorum* show as symptoms poor growth, stunting, wilting and foliage collapse up to plant death. Differences like the order in which foliage wilts and collapses or internal discoloration of the crown tissue are in general used for field diagnosis (Koike and Gordon, 2015) making a good training course necessary to differentiate them, however with some uncertainty.

Soil-borne diseases, like Verticillium wilt, were routinely controlled by soil fumigation with methyl bromide, a very environmentally hazardous chemical forbidden since 2005. Chloropicrin and 1,3-dichloropropene were used to replace it, however nowadays they are not in use anymore. Chemicals authorized as soil fumigants in the EU but not in Germany, included in Annex I, are dazomet and metham as pre-planting application every third year (Colla et al., 2012). Dazomet is the granular and metam sodium the liquid formulation of methyl isothiocyanate. It is a strong herbicide but it does not move through soil like methyl bromide, and can have phytotoxicity problems and inconsistent control when not evenly distributed (Martin, 2003). For other soil pathogens, in conventional production, chemical treatment is allowed: *P. catorum* and *P. fragariae* can be controlled with the active substance Fosetyl-Al alone or in combination with Fenamidon by root dipping before planting and irrigation. Copper oxychloride and hydroxide and benzoic acid are also permitted. Against Verticillium wilt, no PPPs are listed at this time (BVL online). More solutions are needed to ensure the health and productivity of strawberry plants (Vestberg et al., 2004).

Beside the application of biopesticides other different methods and systems can be used to contain soil pathogens in the IPM approach. Most of the methods are used to reduce pest population then to eradicate it (Lamichhane et al., 2017). Beside the use of bio and synthetic pesticides, other options are available mainly as pre-planting controls. For example, soil solarization mechanism, a method useful in lands with high solar incidence, involvs thermal killing of pests along with chemical and biological mechanisms, which play an important role in the lethal process (Gamliel, 2010). In biofumigation, *Brassicaceae* crops products are used as green manures, meals or liquid formulations to suppress pathogen and pests. The effect is achieved by volatile substances resulting from the degradation of plant secondary metabolites and other biological mechanisms (Klosterman et al., 2009; Yim et al., 2016). In past years, the anaerobic soil disinfestation as alternative to chemical fumigants emerged. A broad efficacy spectrum across a range of environments and production systems was shown by the anaerobic decomposition of organic matter (Shennan et al., 2014). In The Netherlands it is used for open fields and uses the incorporation of grass in the soil, watering overhead and coverage with plastic with low oxygen permeability for at least 10 weeks (Goud et al., 2004). The carbon source, soil temperature and the coverage time must be optimized for different production systems to maintain the anaerobic conditions. Volatile compounds released, and the biocontrol by micro-organisms that thrive during the process, seem to be important mechanisms for the suppression of the different pests (Shennan et al., 2014). Crop rotation is an old method used to increase soil fertility and yield and it is a proven IPM strategy to manage soil-borne disease when rotation is made with non host plants (Koike and Gordon, 2015). For example against *Verticillium* MS successful reduction in soil was achieved by rotating plantings with broccoli (Ikeda et al., 2015; Shetty et al., 2000) also for strawberries (Martin and Bull, 2002). To manage Verticillium but also Fusarium wilt and *Phytophthora*, planting disease-free plants when the soil is still free from/of the disease is also a good method (Eikemo et al., 2003; Koike and Gordon, 2015). To plant resistant varieties is also an option (Klosterman et al., 2009) but most cultivars do not have other qualities essential in commercial strawberry production such as flavour, colour, size and shelf life (Eikemo et al., 2003). To follow basic hygienic rules is also an important method to prevent dispensation of harmful organisms. For example, avoiding to move the equipment from known infested to not infested fields, and to wash tractors and farm implements after work can help to reduce spread of the pathogen (Koike and Gordon, 2015).

OBJECTIVE OF THIS THESIS

Development of adequate production, formulation and application methods of microbials are important steps in their commercialization. The efficacy of microbials is a concern, since most studies were made in controlled environments and without formulation resulting in BCAs that show inconsistency in their performance, reducing the users' confidence in these products. Cultivation methods influenced the formulated micro-organism and the formulation influences the BCA storage ability and efficacy. One aim of this thesis is to contribute to the knowledge of the influence of the freeze-drying process on the behaviour of *Pseudomonas* and to see if the fermentation conditions have influence on survival and efficacy of freeze-dried cells of a selected *P. fluorescens* strain. Secondly, the thesis aims to provide information about the applicability of formulated microbial products in the contest of integrated management and commercial conditions.

Optimization of a freeze-drying process for Pseudomonas *spp. and its influence on viability, storability and efficacy* describes the steps used to investigate the practicality of freeze-drying to formulate different *Pseudomonas* sp. To improve the viability, three different freezing rates, three drying temperatures and 21 CPAs were compared. After optimizing the process the efficacy of fresh and freeze-dried cells was tested in a bioassay. The results demonstrate that after optimization of the freeze-drying process the viability and storability of *Pseudomonas* can be increased **(Chapter 2)**.

Influence of different growth conditions on the survival and the efficacy of freeze-dried Pseudomonas fluorescens *Pf153 cells* focuses on the fermentation process. In particular, the growing parameters fermentation time, temperature and media, mild heat shock and pH change on the survival during freeze-drying were investigated. The efficacy of freeze-dried Pf153 cell was tested in a bioassay against *B. cinerea* on *Vicia faba* leaves. The results show, that the quality of the formulated product can be improved by manipulating the fermentation process (**Chapter 3**).

Influence of different temperatures and culture ages on storage survival and efficacy of Pseudomonas fluorescens *Pf153 freeze-dried cells* builds on the results of chapter 3 by studying, in particular, cultivation temperature and harvesting time. Cells grown in selected temperature/time combinations were freeze-dried and stored at 25 °C for 12 weeks. Following freeze-drying, cells were tested in two assay systems against different *B. cinerea* strain. The results show influences on storability but not on the efficacy, however the latter depended on the *B. cinerea* strain and test methods (**Chapter 4**).

Greenhouse and field assessment on the influence of RhizoVital® 42 fl., Trichostar® _and_ Metarhizium brunneum _Ma43 on strawberries in the presence of soil-borne diseases_ describes tests made with commercially available products in different conditions on strawberries. After screening of 68 bacteria and 26 fungi against Verticillium wilt pathogens and _P. cactorum_ in _in vitro_ tests, _Trichoderma harzianum_ T58, _B. amyloliquefaciens_ FZB42 and _M. brunneum_ Ma43 were selected for the presented tests. Greenhouse and field experiments were performed with single or mixed antagonists. The results show, that the applied bioproducts had different performances in different environments. In the commercial field experiments, the results were similar in two consecutive years. In addition, the selected micro-organisms proved their applicability in IPM by testing their compatibility with chemical pesticides (**Chapter 5**).

CHAPTER 2

Optimization of a freeze-drying process for *Pseudomonas* spp. and its influence on viability, storability and efficacy

published as:
Dietrich Stephan, Ana-Paula Matos da Silva and Isabella L. Bisutti (2016): Optimization of a freeze-drying process for the biocontrol agent *Pseudomonas* spp. and its influence on viability, storability and efficacy. *Biological Control*, 94: 74-81 (see Supplement 1).

Pseudomonas are well known ubiquitous bacteria antagonistic to several plant diseases. Despite their abilities, these Gram-negative, non-sporulating, desiccation sensitive bacteria are difficult to preserve. As a matter of fact, limited information about the influence of the formulation process on the viability and efficacy of cells of *Pseudomonas* are available. Here, the practicability of freeze-drying, in general used to conserve desiccation sensitive micro-organism, to formulate and stabilize pseudomonades was investigated. Five different *Pseudomonas* strains were used to test the influence of freezing rates, drying temperatures and the presence of CPA on cells viability after the process (table 3).

Table 3: List of the tests performed on different *Pseudomonas* strains to optimize the freeze-drying process.

Pseudomonas	Comparison of strains[1]	Freezing rates[2]	Drying T[3]	Storage[4]	Bioassay against/on[5]
putida I112	x	x		x	
chlororaphis PCL1391	x	x	x	x	
protegens CHA0	x	x		x	
fluorescens Pf153	x	x	x	x	*B. cinerea/ Vicia faba* leaves
putida MF416		x			*Alternaria dauci* and *A. radicina* /carrots seed

[1]Comparison of strains was made by freezing at 1.3–1.9 °C/min and drying at 5 °C
[2]Freezing rates: 0.6–12 °C/s; 1.3–1.9 °C/min; 0.6–0.8 °C/min; 0.04–0.12 °C/min in the presence of skimmed milk as CPA
[3]Drying temperatures (T): 5, 20, 30 °C after freezing at 1.3–1.9 °C/min in the presence of skimmed milk as CPA
[4]Storage: at 40 °C up to 7 days after freezing (using the best rate for each strain) and drying at 30 °C
[5]Bioassay with fresh and freeze-dried cells

The fermentations of the bacteria were performed in flasks incubated on a horizontal shaker at 150 rpm to the stationary growth phase. Figure 3 describes the steps from the cultivation to the preparation of the sample for the followed process. Freezing and drying were performed in a freeze-drier except for the shock freezing that was done with liquid nitrogen. The other freezing rates were obtained with computer supported cooling from 5 to -40 °C over 40, 50 min or 16.7 h. The number of living cells was estimated by the most probable number (MPN) method before and after freeze-drying. The freeze-dried samples were re-suspended with water to the original weight before assessment.

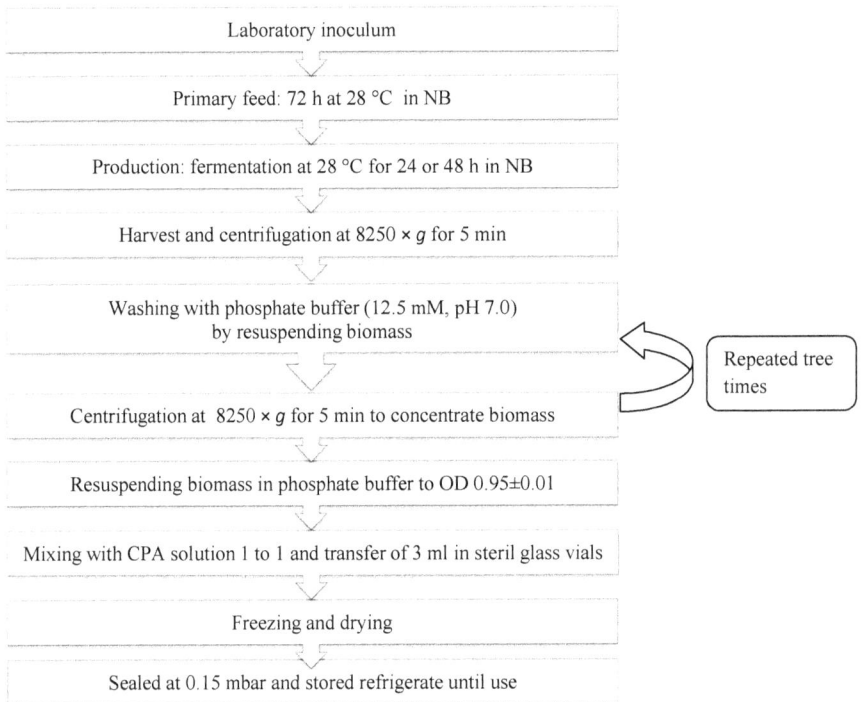

Figure 3: Steps followed during production and formulation of *Pseudomonas* strains used in the tests described in table 3

The results show, that the viability of four strains after freeze-drying ranged only between 2% and 10% when the protocol was not optimized. The freezing rate did not influence significantly the viability after the process. However, for each strain a slightly better survival was freezing rate depended and changed for the five tested strains. An increase in drying temperature showed also a better survival of the tested strains, but for *P. chlororaphis* PCL1391 these were statistically lower than before drying. With Pf153 it was

40

shown that CPAs have a particular influence on the viability after freeze-drying. Therefore, 20 different CPAs were used during freezing (1.3–1.9 °C/min) and drying at 5 °C for 18 h. The viability was the highest for saccharose followed by lactose, ligninosulfonic acid, glucose and skimmed milk. Glycerol, bentonite, activated carbon and alkalic lignin reduced the viability to values lower than when cells were freeze-dried alone. In storability tests, lactose and skimmed milk showed their potential to protect vegetative cells during a longer period. The efficacy of *P. putida* of MF416 freeze-dried cells did not differ from freshly produced ones. The same result was obtained for Pf153 in the presence of lactose or skimmed milk in another plant-pathogen system, however when Pf153 was applied formulated in saccharose, its efficacy against *B. cinerea* decreased dramatically.

The results demonstrate, that for pseudomonades freeze-drying is an interesting conservation technique. When optimizing the freeze-drying process, different criteria, such as viability, efficacy and storability have to be considered. Within the protocol, the selection of the right protectant is important because it influences viability, storability and the efficacy.

CHAPTER 3

Influence of different growth conditions on the survival and the efficacy of freeze-dried *Pseudomonas fluorescens* Pf153 cells

published as:
Isabella L. Bisutti, Katia Hirt and Dietrich Stephan (2015): Influence of different growth conditions on the survival and the efficacy of freeze-dried *Pseudomonas fluorescens* strain Pf153. *Biocontrol Science and Technology*, 25 (11): 1296-1284 (see Supplement 2).

During the development of a commercial product containing microbial BCAs, production and formulation are important steps to be considered. Indeed, by manipulating the fermentation process, it is possible to improve the quality of a formulated product as shown for Pf153. Therefore, the influence of different cultivation parameters on the survival after formulation and on the biocontrol efficacy were investigated. The fermentation parameters studied were: harvesting time, growth temperature and media, and stresses as mild heat shock and pH change. The bacterium was pre-cultured in 30 ml KB in flasks for 24 h at 28 °C with 150 rpm before inoculating it with 1:100 of the main culture. To determine the influence of the fermentation time, a sample was taken every four hours from a fermenter (figure 4) and optical density (OD) at 595 nm was determined. For the experiments on the influence of media composition, growth temperature and heat shock, flaks shaken at 150 rpm for 16 h were used. The tested media differed in carbon and nitrogen sources and ratios. Heat shocks were applied after 15 h fermentation at 28 °C by increasing the temperature for one hour. The pH change experiments were performed in

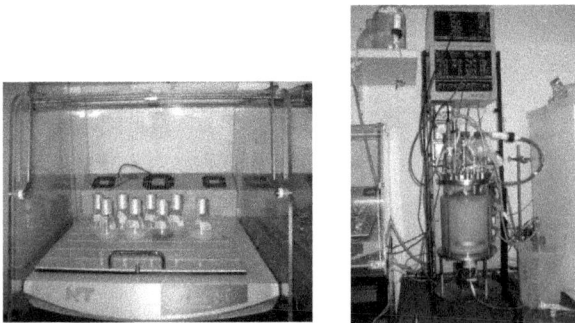

Figure 4: Flasks on shaker (left) and ISF 100 fermenter (Infors, CH) containing 3.0 l KB medium and kept at 28 °C, with 60 NL/h aeration and an agitation of 900 rpm (right).

the fermenter and after 15 h fermentation, the pH was changed of one unit every 30 min by addition of HCl (pH range from 7 to 4) or NaOH (pH range from 7 to 10). The formulation consisted in freeze-drying the cleaned bacterial cell (except for fermentation time trials) in the presence of 10% (w/v) lactose as CPA with an optimized protocol (freezing to -40 °C at rate 0.04-0.12 °C/min and drying at -20 °C, 0.15 mbar for 18 h). The number of living cells was estimated by the MPN method before and after freeze-drying. The freeze-dried samples were re-suspended with water to the original weight before assessment.

As expected, the biomass increased over time up to $2.5x10^{10}$ MPN/ml after 24 h fermentation, however not the survival after freeze-drying. The highest survival was shown for 16 and 20 h fermentation (table 4). For the following experiments, 16 h was chosen as harvest time. Survival rates between 28 and 100% were found. KB medium showed the highest survival rate between media. Growth temperatures of 25 and 30 °C and a mild heat shock at 35 °C for one hour influenced the survival rate positively. pH changes between 6 and 9 did not significantly influence the survival rate, however the extremes 4 and 10 reduced the number of fresh living cells and survival. The efficacy was tested on detached *V. faba* leaves inoculated with *B. cinerea* (description in figure 5) and all tested parameters showed significant efficacy increase compared to the untreated control. No significant

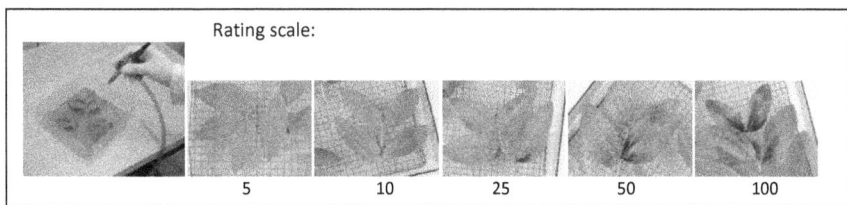

Figure 5: Biotest on *V. faba* leaves. Pf153 treatments ($5x10^8$ MPN/ml) and *B. cinerea* ($5x10^5$ conidia/ml) were sprayed on the leaves by air-brush (left). After five days at 20 °C with 16/8 h day/night cycle, rating scale was used to assess the leaf necrosis. With the equation $\frac{\sum(rating\ number \times number\ of\ leaves\ in\ the\ rating)}{total\ number\ of\ leaves \times highest\ rating} \times 100$ the percentage of affected leaf area was calculated.

differences between fresh and freeze-dried cells of Pf153, produced in different media were shown. Also, the formulated cells treated with one hour heat shock did not show significant differences in efficacy between treatments. Significant differences in efficacy were obtained for Pf153 cultivated at different temperatures. Best results and significantly higher efficacy respective to the other treatments was shown for Pf153 produced at 20 °C (table 4).

For Pf153 freeze-drying is an efficient conservation method with high survival, efficacy comparable to fresh cells and an easy to handle product. To enhance survival during the freeze-drying process, fermentation conditions have to be optimized. These also can have an influence on the performance of the resulting product.

Table 4: Tested fermentation parameters on their influence on the survival of Pf153 after freeze-drying.

Media	Growth temperature °C	Heat shock °C	pH change
TSB ½(52)	20 (40)	28 (44)	4 (38)
DF (62)	25 (99)	35 (103)	5 (61)
KB (73)	30 (85)	40 (69)	6 (71)
	37 (28)	45 (61)	7 (136)
			8 (161)
			9 (93)
			10 (33)

First number is the parameter followed by (% survival rate)
The tests were made in KB medium (except the media trial) and by fermenting at 28 °C (except the growth temperature trial) in flasks except for the pH change, which was made in a fermenter

CHAPTER 4

Influence of different temperatures and culture ages on storage survival and efficacy of *Pseudomonas fluorescens* Pf153 freeze-dried cells

submitted to the *Journal of Applied Microbiology* after editing editor and reviewers comments as:
Isabella L. Bisutti and Dietrich Stephan (2019): Influence of fermentation temperature and duration on survival and biocontrol efficacy of *Pseudomonas fluorescens* Pf153 freeze-dried cells (see Supplement 3).

Fermentation and formulation are important steps during production of biopesticides. They affect efficacy and storage ability of the biocontrol agent. Based on previous results, where Pf153 showed an increased efficacy against *B. cinerea* on *V. faba* leaves when fermented at 20 °C respective to higher temperatures, the effect of fermentation temperature and time on the storage and efficacy of freeze-dried cells was assessed.

The fermentation and formulation procedure is described in figure 6. The tested temperatures were 20 °C and 28 °C and by means of OD measurement, the harvesting times were selected. Cells were collected in the middle of the logarithmic phase (after 8 h for 28 °C and 16 h for 20 °C) and at the beginning of the stationary phase (after 16 h for 28 °C and 28 h for 20 °C). The formulated cells were stored at 25 °C for up to 12 weeks and the number of living cells was estimated by the MPN. For the freeze-dried cells, the samples were re-suspended with water to the original weight before assessment. The efficacy was tested against different *B. cinerea* strains in two test systems: on *V. faba* leaves and with a modified dual culture tests on two different media.

As expected, the MPN/ml were higher at the beginning of the stationary phase compared to the middle of the log phase. The survival after freeze was higher for cell growth at 20 °C and younger cells at 28 °C. Freeze-dried cells fermented at 20 °C showed a better storage survival after 5 weeks than those grown at 28 °C irrespective to the fermentation time. However, the viability was reduced over time from about 10^9 to $10^7/10^6$ MPN/ml or less for all fermentation parameter combinations.

Depending on assay form and *B. cinerea* strain, different results in the efficacy tests were obtained. Tested on *V. faba* leaves, the temperature/time combination 28 °C/ 8 h had a significantly better efficacy than the other combinations. In the *in vitro* test against different *B. cinerea* strains on two different media none of the diverse temperature/harvest time combination excelled over the others. Except for one *B. cinerea* strain their efficacy was significantly higher than the control. When considering all results of the *in vitro* tests,

Pf153 fermented at 20 °C for 28 h was the best in reducing mycelial growth in six of ten tests and the second best in the remaining four. Even though no one temperature/harvest time combination was the best in all the considered criteria, Pf153 fermented at 20 °C for 28 h showed the best combination between biomass production, survival after formulation, storage and *in vitro* efficacy. In any case, it is important to test the efficacy of the end product against more strains of the considered pathogen and in different assays.

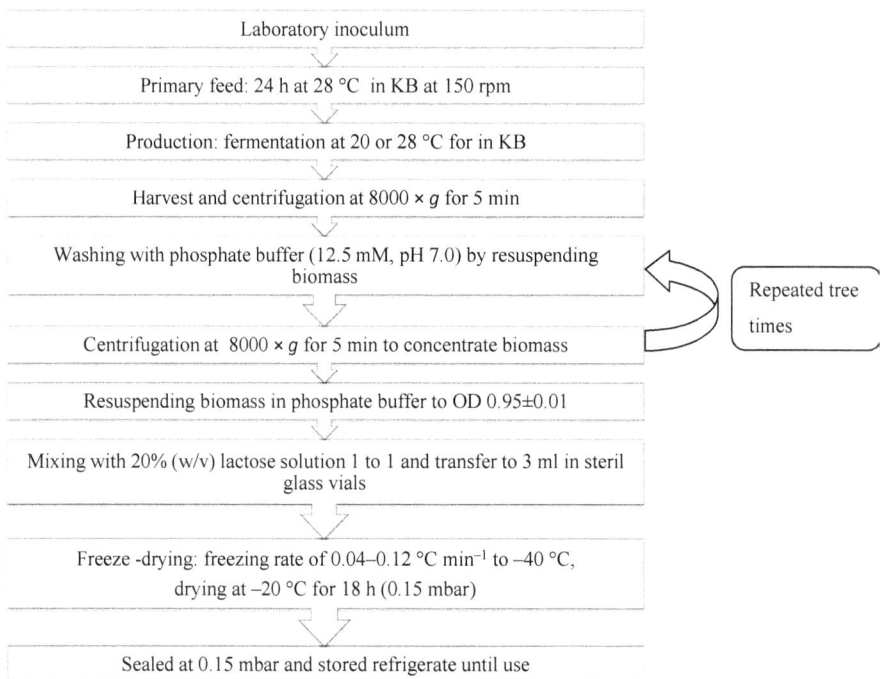

Laboratory inoculum

Primary feed: 24 h at 28 °C in KB at 150 rpm

Production: fermentation at 20 or 28 °C for in KB

Harvest and centrifugation at 8000 × g for 5 min

Washing with phosphate buffer (12.5 mM, pH 7.0) by resuspending biomass

Repeated tree times

Centrifugation at 8000 × g for 5 min to concentrate biomass

Resuspending biomass in phosphate buffer to OD 0.95±0.01

Mixing with 20% (w/v) lactose solution 1 to 1 and transfer to 3 ml in steril glass vials

Freeze -drying: freezing rate of 0.04–0.12 °C min^{-1} to –40 °C, drying at –20 °C for 18 h (0.15 mbar)

Sealed at 0.15 mbar and stored refrigerate until use

Figure 6: Steps followed to produce formulated Pf153.

CHAPTER 5

Greenhouse and field assessment of the influence of RhizoVital® 42 fl., Trichostar® and *Metarhizium brunneum* Ma43 on strawberries in the presence of soil-borne diseases

published as:
Isabella L. Bisutti, Juliana Pelz, Carmen Büttner and Dietrich Stephan (2017): Field assessment on the influence of RhizoVital® 42 fl. and Trichostar® on strawberries in the presence of soil-borne diseases. *Crop Protection* 96: 195-203 (see Supplement 4).
and internal report:
Isabella L. Bisutti and Dietrich Stephan (2014): Einsatz mikrobiologischer Präparate zur Regulierung der bodenbürtigen Erdbeerkrankheiten *Verticillium dahliae* und *Phytophthora cactorum* sowie des Erdbeerblütenstechers (Appendix to supplement 4: Isabella L. Bisutti (2017): Selected micro-organisms in greenhouse and controlled field trials against *Verticillium dahliae* and *Phytophthora cactorum* on strawberries. Extract of "Material and methods and Results" of final project report FKZ 2811NA012).

High losses in yield of strawberries, a beloved fruit in Germany, are caused by diverse diseases and arthropod pests. The new EU regulation that promotes IPM management with the reduction of chemical pesticides application, could increase the disease problems in areas were chemicals were the only used control.

In the study presented, the two commercially available soil strengtheners RhizoVital® 42 fl. and Trichostar®, and the fungus *M. brunneum* Ma43, used against strawberry blossom weevil, were tested for their compatibility with chemical pesticides and used in diverse trials to evaluate their ability against two soil-borne pathogens. The trials were performed in the greenhouse and in controlled and commercial fields (figure 7).

Figure 7: Commercial field with confirmed soil pathogens (left) and commercial field at strawberries ripening beginning.

The compatibility with 14 chemical pesticides was tested *in vitro* by combining the bio with the chemical pesticide for up to 4 h while counting the colony forming unit (CFU) for the bacteria and the germination capacity for the fungi after the contact. The CFU of *B. amyloliquefaciens* FZB42 contained in RhizoVital® 42 fl. were not reduced under the declared concentration by any tested synthetic pesticide used in this study at any contact time. For *T. harzianum* T58 contained in Trichostar® only one reduced the percentage of germinated conidia to almost zero. On the other hand, fungicide of the oximinoacetate group increased germinated conidia at least 10% relative to the control. No significant

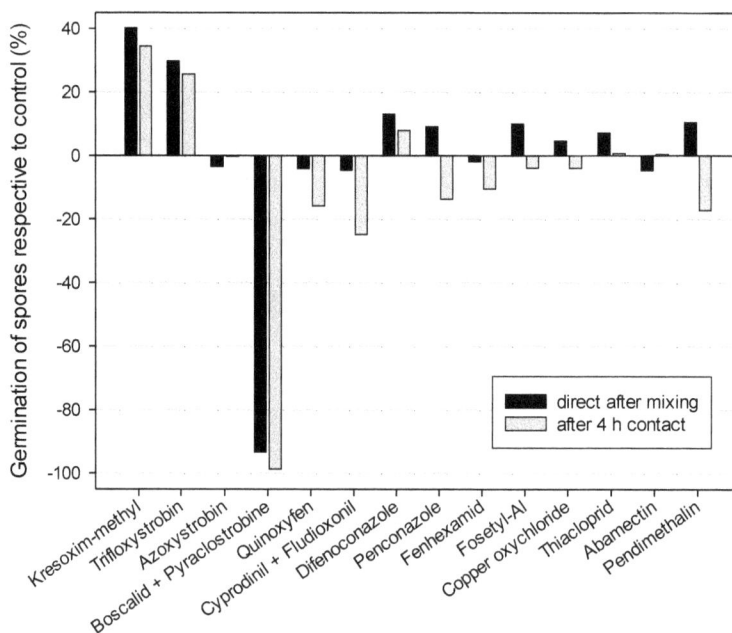

Figure 8: Influence of chemical pesticides on the germination of Trichostar® spores direct after mixing (black columns) with pesticides or after four hours contact (gray columns). Values are reported in percentage relative to the germination of spores without chemical pesticide contact (untreated control is reported as zero line).

differences were shown between the tested contact times (figure 8). At least 80% of the spores of *M. brunneum* Ma43 germinated after mixing with seven of the 13 tested pesticides at any contact time. All the fungicide influencing respiration reduced the percentage of germinated spores markedly, although less after 4 h contact than directly after mixing. Signum® reduced the number of germinated spores to almost zero as for *T. harzianum* T58 (figure 9).

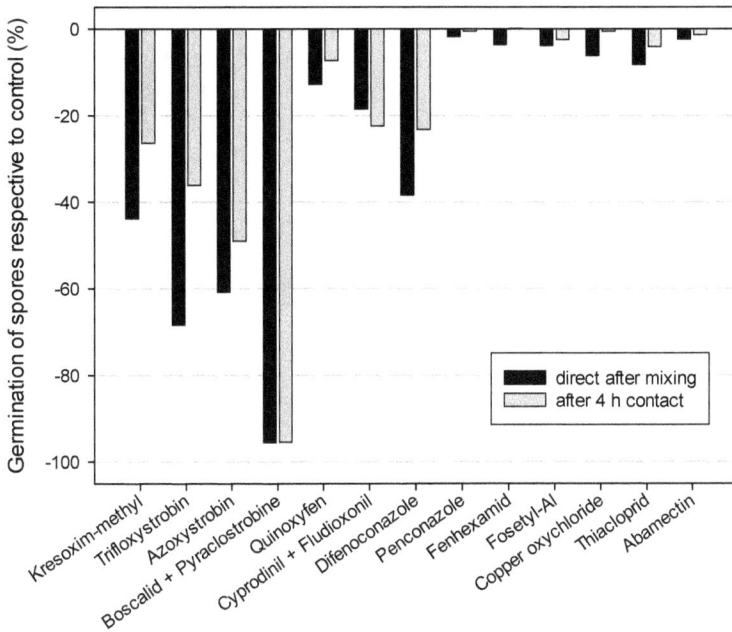

Figure 9: Influence of chemical pesticides on the germination of *M. brunneum* Ma43 spores direct after mixing (black columns) with pesticides or after four hours contact (gray columns). Values are reported in percentage relative to the germination of spores without chemical pesticide contact (untreated control is reported as zero line).

The trials against *V. dahliae* and *P. cactorum* were performed by adding the pathogens to the potting soil or in the planting hole before planting the strawberries after dipping them into the micro-organism suspensions (figure 10). For the commercial field trials, no artificial inoculation was made. The results of the greenhouse trials showed, that for plants treated with Trichostar® in the presence of *P. cactorum* the considered growth parameters (plant fresh and dry weight, new roots and areal part dry weight, length of the longest root, number of green leaves, number and dry weight of daughter plants) were mainly lower than the control. On the other hand, for the application without disease inoculation, the parameters were higher than the control. Also the other treatments did not excel in their performance. Plants inoculated with *V. dahliae* did not show symptoms, whereas for *P. cactorum* almost 30% of the plants died (the most in the untreated control). In the controlled field the result did not show any positive effect on yield or plant vigour by inoculating the strawberry plants with the soil strengthener or/and *M. brunneum* Ma43 neither for the pathogen inoculated plants nor for the not pathogen inoculated plants. This could be caused by the fact, that before starting the trials, compost preparation was added

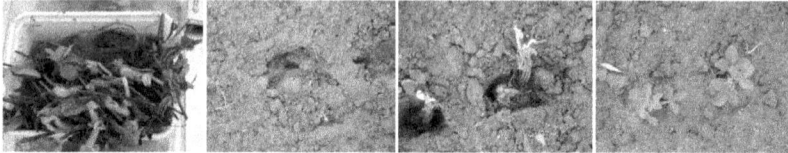

Figure 10: Planting sequence used for the greenhouse and field experiments (with specific variation for each environment). Dipping of the plants in the antagonist for 15 min before planting; addition of the pathogen to the soil (not in commercial fields); planting of the strawberries and ca. 15 days after planting (from the left to the right).

to the field to increase fertility. It can be supposed that the compost added a conspicuous number and variety of microbials that acted adverse to the inoculated antagonists. In parallel to the controlled trials, RhizoVital® 42 fl. and Trichostar® were tested in two commercial fields in the Rhine-Main area known to contain Verticillium wilt pathogens producing MS. In one commercial strawberry field, Trichostar® with just one treatment in 2013, increased yield of 8 and 9% in 2013 and 2014 respectively. In the second field, RhizoVital® 42 fl. increased yield about 6% each year after application of the soil strengthener in both years (table 5).

Table 5: Total yield of the two commercial fields inoculated with soil strengtheners by root dipping.

	Field 1		Field 2	
	Yield* 2013	Yield 2014	Yield 2013	Yield 2014
Control	17.1	21.9	36.9	78.6
RhizoVital® 42 fl.	20.8 (21.3%)	22.5 (2.7%)	38.9 (**5.6%**)	83.7 (**6.5%**)
Trichostar®	18.5 (**7.9%**)	23.9 (**9.2%**)	40.5 (9.7%)	73.9 (-6.0%)
Mixture	19.6 (14.3%)	21.2 (-3.0%)	36.4 (-1.3%)	80.7 (2.7%)

The two fields were situated in the Rhine-Main region, with different soil composition and farming method
*weight in Kg (increase respective to the untreated control)

The results show, that the bioproducts and *M. brunneum* Ma43 could be integrated in a treatment schedule. Application of micro-organisms to strawberry plants can improve growth and yield, but the selection of the right product for the right environment is of crucial importance.

CHAPTER 6

GENERAL DISCUSSION

Bisutti Isabella L.

Organic product demands are increasing; indeed, the food industry's request of organic foods and beverages is a rapidly growing segment. A reason why more people buy organic food is, that they believe these products are more nutritious and that organic farms produce better tasting food from healthier soils. For example, it was found that organically produced strawberries are of higher quality with a longer shelf life (Reganold et al., 2010). Strawberries are a popular "fruit" that has gained high recognition among consumers for its particular taste and health benefits. Its demand is also great in the food industry, due to its versatile application as a flavouring agent or in fruit preparations (Bhat et al., 2015). Strawberry production in Germany is lower than the demand, which makes import a necessity to satisfy the market (BZFE, 2017). In a case study from 2015, consumers showed a preference for strawberries that were locally grown. Street vending booths were very popular and people perceived berries sold there as "more fresh" and coming directly from the farm. Even though in this survey 42% of the participants had no preference regarding farming practices, 35% preferred organic versus nearly 21%, which preferred conventionally grown strawberries (Bhat et al., 2015).

Strawberries are counted among the fruits that are heavily contaminated by chemical pesticide residues (BZFE, 2017). Crop protection depended on synthetic chemical pesticides for decades, but their use is generally becoming significantly more difficult due to new health and safety legislation. The indiscriminate use and prophylactic misuse of pesticides can damage human health and the environment, and can result in management failure and resistance development. Furthermore, concerns about the safety of pesticide residues in food are expressed by consumers and pressure groups (Chandler et al., 2011; Tiwari and Tripathi, 2014). Organic farming can deliver products with lower or no pesticide residues. Organic systems however are considered an inefficient approach to food production because yield is lower compared to conventional agriculture (Muller et al., 2017; Reganold and Wachter, 2016). On the other hand, when environmental impact, economic viability and social wellbeing are considered, organic is one step ahead because it is more profitable, environmentally friendly and the products contain less or no pesticide residues while being equally, if not more nutritious foods.

A blend of organic and alternative tactics are required to make crop protection more sustainable (Reganold and Wachter, 2016) and IPM is promoted by many experts. IPM includes the combination of different crop protection practices and one measure includes the biological control of plant pathogens with microbial antagonists which are part of biopesticides (Chandler et al., 2011; Tiwari and Tripathi, 2014). Reliance on BCAs is considered not effective due to the often limited efficacy. Combined with other IPM methods, however, they are relevant to reduce the dependence on conventional pesticides (Lamichhane et al., 2017). In model systems, numerous micro-organisms showing an antagonistic ability can be found because the efficacy is generally the main criterion for most screening programs. However, forgetting characteristic relevant for commercialization reduces the number of antagonists that meet the requirements for commercial use (Köhl et al., 2011; Yang et al., 2014). Since biological control products based on micro-organisms are considered as PPPs, they undergo the same government regulations for registration designed originally for synthetic chemical pesticides (Chandler et al., 2011; Köhl et al., 2011). Beside the government requirements, there are also significant technical barriers to make biopesticides more effective. This challenge opens new opportunities when developing biopesticides by combining ecological science with post-genomics technologies (Chandler et al., 2011). Therefore, during screening programs for candidates suitable for commercial use, government requirements and quality improvements have to be additionally considered with disease control (Köhl et al., 2011; Yang et al., 2014).

Freeze-drying *Pseudomonas*

For different beneficial *Pseudomonas* the positive influence on plants are well described. *Pseudomonas* can improve plant growth and yield directly and indirectly. They can act as plant protection bacteria by producing diverse groups of metabolites, through competition for iron and other nutrients, by niche exclusion, by parasitism and by inducing systemic resistance (Jain and Das, 2016; Nagarajkumar et al., 2004). Even though *Pseudomonas* have all these useful and positive traits, and are considered interesting potential BCAs, they are still difficult to formulate. To use freshly fermented liquid cell suspension as inoculant under field conditions is rather uncommon (Gade et al., 2014; Shen et al., 2016). Hence, it is useful to immobilize the cells of *P. fluorescens* in certain carriers to prepare easy applicable, storage, commercialization and field use formulations (Gade et al., 2014). The selection of the conservation modality is very important. *P. fluorescens* are Gram-negative bacteria that do not produce spores and freeze-drying is up to now the best method to obtain a powder with high viability of cells (Mputu, 2014). Appearance of the end product and viability of the micro-organism can be affected by diverse parameters within the process (Morgan and Vesey, 2009; Morgan et al., 2006). Part of this thesis focuses on the optimization of diverse parameters to increase viability, storability and efficacy of different *Pseudomonas* strains after freeze-drying (chapters 2, 3 and 4).

Freeze-drying process

During **freezing**, the water forms ice crystals and the solute is trapped in between. These interstitial matrix components maintain the solid structure during removal of the frozen water (Morgan and Vesey, 2009). Different freezing rates affect the viability of the micro-organism: in the intracellular compartment ice nucleation and crystal growth may occur, or it may vitrify (Fonseca et al., 2016). Slow controlled freezing is considered to prevent damage of the membranes (Morgan and Vesey, 2009) but the exposure to increasing solute concentrations can stress the cell and reduce survival (Zhao et al., 2016). A rapid freeze,

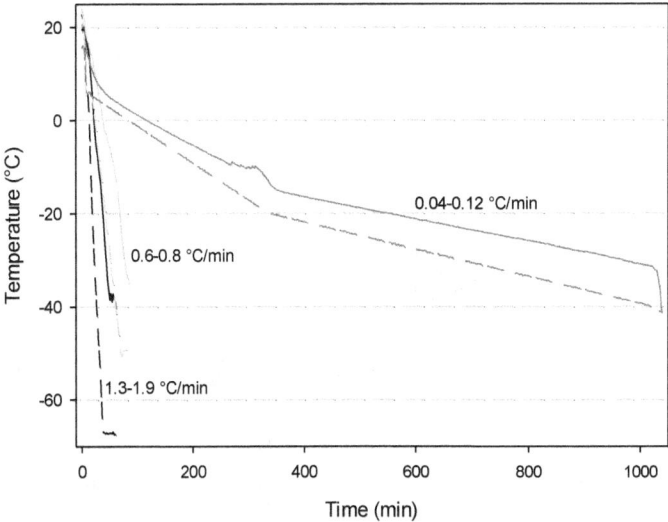

Figure 11: Product (dash line) and shelf temperature (solid line) during the freezing process at different freezing rates. Mean of three independent experiments. For the freezing rate 1.3-1.9 °C/min the shelf of the freeze-dryer was kept for 10 min at 5 °C and afterwards held at -70 °C for 50 min; for the freezing rate 0.6-0.8 °C/min the shelf was kept 20 min at 5 °C and then the temperature was reduced in 50 min at -50 °C, for the slowest freezing rate of 0.04-0.12 °C/min after the shelf was kept for 20 min at 5 °C, the temperature was reduced in two steps, the first at -20 °C in 315 min and the second to -40 °C in 685 min.

like with liquid nitrogen, is considered to minimize damage caused by crystal growth (Morgan and Vesey, 2009), however the formation of intracellular ice could be a problem (Zhao et al., 2016). In this thesis, different freezing rates were tested. The bacterial suspension temperature was brought to -40 °C in a freeze-dryer by different instrument settings (figure 11). The shock freezing was obtained by addition of liquid nitrogen. The tested *Pseudomonas* strains did not show significant differences in viability between freezing rates, however differences between strains were observed. *P. protegens* CHA0 showed the best resistance to freezing all over. For *P. putida* MF416 and *P. chlororaphis* PCL1391 the differences in survival between rates were the smallest, with strain PCL1391 having a better freezing resistance. *P. putida* 1112 and Pf153 showed to be the most influenced by the freezing rates (chapter 2). These results confirm, that the resistance of bacteria to freezing depends on the freezing conditions and on the strain (Fonseca et al., 2006; Mputu et al., 2012b) in addition to the culture conditions, the harvest time and the formulation considering type and concentration of protectans (Fonseca et al., 2006).

Stabilization of the material is achieved by reducing the solvent by **drying** (Morgan and Vesey, 2009), but after the process the mortality can reach values of more than 90% (Bashan et al., 2014; Schisler et al., 2016). At the beginning, the frozen water is removed by sublimation (primary drying) followed by removal by desorption (secondary drying). Sublimation temperatures have to be calibrated to avoid collapse and melting of the product. This is achieved by low shelf temperatures applied for a longer time and reduction of the ambient pressure below ice vapour pressure of the product. After sublimation, the adsorbed water is confined within the solid matrix and its removal takes extended periods of time (Morgan and Vesey, 2009; Morgan et al., 2006). The tested drying temperatures reported in this thesis, were applied by setting the shelf at 5, 20 or 30 °C with a chamber pressure of 0.2 mbar. The temperature profiles are presented in figure 12. For both tested *Pseudomonas* strains, the highest drying temperature showed the highest survival, however survival of Pf153 was almost 20% higher than *P. chlororaphis* PCL1391 (chapter 2). For antagonistic fungi, where the conidia are used, the rate of drying is known to be a factor that influences viability during the process and storage. Gradual and slow drying seems to have better conidia survival and viability than rapid drying (Daryaei et al., 2016c). Drying remains a challenging and critically important step when developing BCAs in commercial products (Schisler et al., 2016). Therefore, when screening for drying protocols, survival rate has to be considered as one of the main quality parameters (Berninger et al., 2018).

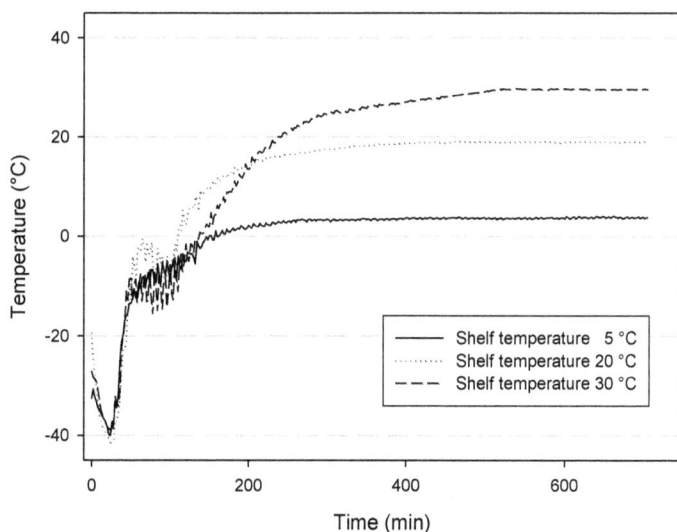

Figure 12: Product temperature during the drying process with different fixed shelf temperatures. The shelf temperature was reached after ca. 670 min for 5 °C, after ca. 620 min at 20 °C and ca. 700 min at 30 °C. The drying at 5 °C shelf temperature was faster than at 30 °C.

Co-formulation substances

The addition of a carrier can increase the number of viable and active micro-organism for field application, supporting the success of a biopesticides (Keswani et al., 2016). Carriers are inert ingredients that do not have biocontrol capabilities but can strongly affect efficacy and shelf life of the end product (Fravel et al., 1998; Keswani et al., 2016; Wei et al., 2015). Carriers can provide a food base to aid proliferation or protect the cells during formulation. Since beneficial micro-organism are considered eco-friendly, it is mandatory that any additives in the formulation should be eco-friendly (Keswani et al., 2016). Survival of *Pseudomonas* cells, after freeze-drying without any CPAs or LPAs, are reported to range between 1 and 10% depending on the strain (Mputu and Thonart, 2013; Stephan et al., 2016). Modifications in the physical state of membrane lipids and in the structure of sensitive proteins, cause a decrease in viability of cells during and after the process (Mputu et al., 2012b). When the bound water molecules are removed, the membrane density and the transition phase temperature can rise, so that rehydration can affect cell viability. Drying in the presence of some sugars lowers this transition temperature (Morgan and Vesey, 2009). Since viability after freeze-drying can vary when using different CPA (Zhao et al., 2016), in this thesis, 20 different substances were tested in their ability to improve viability of Pf153 after the process. They were added before freezing and were chosen in diverse groups including carbohydrates, proteins and polymers. These additives are glass-forming substances that have been shown to exert the highest protection during the process (Morgan and Vesey, 2009). The best are carbohydrates (Muñoz-Rojas et al., 2006; Stephan et al., 2016) which can replace hydrogen bonds, lower membrane transition phase, preserve proteins or provide structural support during freeze-drying (Morgan and Vesey, 2009). For *P. fluorescens* BTP1 it was shown, that freeze-drying without CPAs affected mostly the polyunsaturated fatty acids (Mputu et al., 2012b) and that a protective effect could be obtained by glycerol and maltodextrine (Mputu and Thonart, 2013). In general, disaccharides are considered the better CPAs for high survival. For example, saccharose can raise the glass-phase transition temperature and stabilize the cytoplasmic membrane properties during freeze-drying (Zhao et al., 2016). Lactose is often reported to improve survival of *Pseudomonas* spp. during freeze-drying (Muñoz-Rojas et al., 2006) by providing structural support and by crystallizing (Morgan and Vesey, 2009). In our tests lactose was the second best CPA considering survival

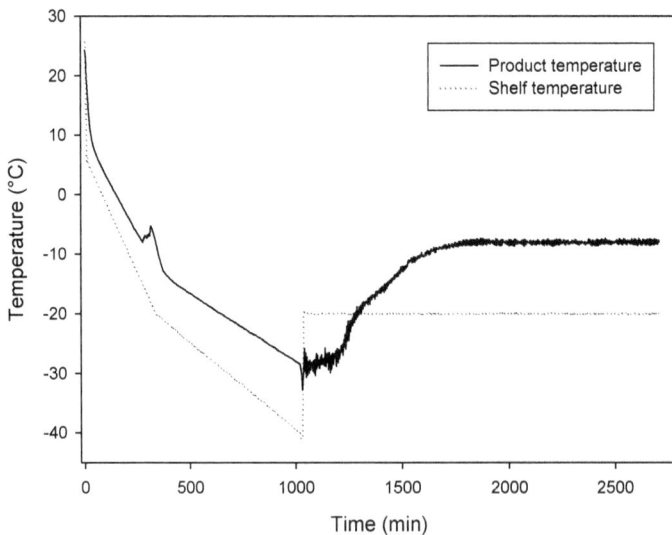

Figure 13: Freeze-drying profile used for the experiments with Pf153.

after freeze-drying however, during drying at 30 °C the cake sometimes melted, making the dissolving of the product not feasible. Therefore, for the following experiments reported here, the drying temperature was reduced to -20 °C with longer drying times (figure 13). These changes can be a reason why the reported survival after freeze-drying was reduced respective to the data reported by Stephan et al. (2016).

The selection of CPA, LPA, carriers and other adjuvants should include their influence on the **efficacy**. Formulation substances are important for the antagonist activity, since they can increase effectiveness significantly (Helbig, 2002), but in some cases they may repress biocontrol activity and consequently they should be avoided (Fravel et al., 1998; Wei et al., 2015). The ideal carrier should not only protect during formulation and prolong shelf life, it should also promote efficacy (Wei et al., 2015). The importance of adjuvants was shown in tests made with the biofungicide Serenade®, which contains *B. subtilis* QST713. There, the individual components were less effective in greenhouse trials in suppressing infection of canola by *Plasmodiophora brassicae* than their combination in the product (Lahlali et al., 2013). Also for *A. quisqualis* AQ ITA3 the addition of diverse carriers sustained the efficacy against *Erysiphe necator* and *Podosphaera xanthii* compared to the unformulated fungus (Angeli et al., 2016). Similar for *B. amyloliquefaciens* QL-18 when formulated in rapeseed cake compost against *Ralstonia solanacearum* (Wei et al., 2015). In my investigations three CPA were tested to determine their influence on efficacy of fresh and freeze-dried cells of Pf153. Skimmed milk increased the *B. cinerea* diseased area on *V. faba* leaves when applied alone, but as part of the formulation helped to increase the efficacy of Pf153, also of the freeze-dried cells. Added to a medium in Petri plates as carbon source (figure 14) it stimulates the sporulation of *B. cinerea*. The positive effect of skimmed milk on efficacy was also shown for *A. quisqualis* AQ ITA3 in liquid formulation against *E. necator* and *P. xanthii*. This positive effect was also reported for the addition of saccharose (Angeli et al., 2016), however in my experiments the addition of saccharose to the cell suspension reduced the ability of Pf153 to control the disease. Reduction of efficacy was also shown for *P. fluorescens* S11P12 where the addition of osmoprotectans during fermentation and carriers during formulation increased the Fusarium dry root lesions on potatoes (Schisler et al., 2016). Therefore, during carrier screening studies, the target pathogen has to be considered to avoid nutrients and substrate supply for its activity (Schisler et al., 2016; Wei et al., 2015).

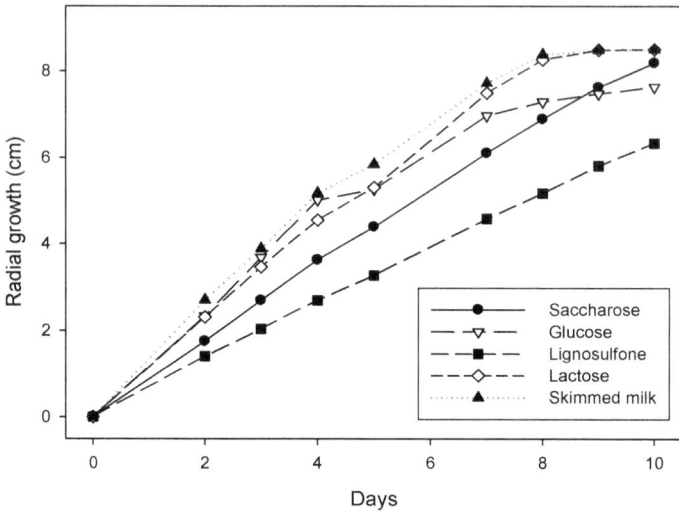

Figure 14: Mycelial growth of *B. cinerea* on agar plates amended with different carbon sources at 20 °C in the dark. The reference medium is the Czepak Dox agar (i.e. (l^{-1}) 30 g saccharose, 3 g NaNO$_3$, 1 g K$_2$HPO$_4$ x 3 H$_2$O, 0.5 g MgSO$_4$ x 7 H$_2$O, 0.5 g KCl, 0.01 g iron (II)-sulfate, 18 g agar-agar) where the carbon source was replaced by glucose, lignosulfone, lactose and skimmed milk.

Storage

A storage period between manufacturing and use is usually required. During storage, the biological traits have to be retained and the number of CFU should be maintained at a certain level. Storage at room temperature for more than one year is often necessary to successfully integrate a product into the agricultural distribution system, especially in developing countries (Bashan et al., 2014). The reduction of the microbial population over time, under determined levels, can be overcome by increasing the initial CFU/ml content or by manipulating production and formulation technology and/or storage conditions like packaging and transport, since shelf life depends on several factors and their interaction (Keswani et al., 2016).

CPA are almost essential for *Pseudomonas* during the freeze-drying process and influence, in combination with the temperature, the storage survival over a period of time (Mputu et al., 2012b). In general, for freeze-dried micro-organisms the maximum viability expected during storage at temperatures higher than 30 °C are 7 days and one year at 4 °C. In this thesis, five CPA were tested for their influence on storage of four *Pseudomonas* strains. All *Pseudomonas* strains tested showed viability higher than 10^8 MPN/ml after one week at 40 °C when formulated in lactose and skimmed milk, being in the expected values reported by Morgan and Vesey (2009). Lignosulfonic acid showed storage abilities for *P. protegens* CHA0 and *P. putida* I112 where the MPN/ml were at least 10^4. Saccharose was only good for Pf153 and glucose failed conservation for all strains tested (chapter 2). Since carrier's performance in maintaining dried cells viable can vary considerably, a strain by strain assay is always needed/necessary (Schisler et al., 2016).

In a longer storage test made at 25 °C for 2 months (Stephan et al., 2007), saccharose and lactose showed again high viability with values of $2.7x10^8$ and $2.2x10^8$ MPN/ml for Pf153 at the end of the test. Pf153 stored with skimmed milk reduced its viability to $1.3x10^7$ MPN/ml, confirming the results obtained by high temperature storage tests. The influence of CPA, but also of the temperature, on shelf life is reported for freeze-dried *Bifidobacterium crudilactis*, which was longer storable at 4 °C than at 23 °C for all CPA tested (Tanimomo et al., 2016). Also for *P. fluorescens* BTP1 lower temperature, storage tests made at 4 °C and at 20 °C, showed better results at low temperature with survival up to 270 days in particular with glycerol (Mputu et al., 2012b). Unfortunately, the best CPA during freeze-drying and the one that assures longer shelf life, are not always the same. When maltodextrine was added as CPA to *P. fluorescens* BTP1, the highest survival after freeze-drying was reached, but not in the storage viability (Mputu et al., 2012b). Also in my tests saccharose, which was the best CPA during freeze-drying of Pf153, was not the best for storage, being the second best behind lactose (chapter 2).

In the literature, reports of storage up to 14 years can be found, however to evaluate storage stability over a longer time slows down the development of new formulations (Berninger et al., 2018). The accelerated shelf life testing method (ASLT) is widely used for the prediction of storage stability and quality, and for the estimation of shelf life and storage temperatures of perishable products (Franks, 1994; Hernández et al., 2006; Mizrahi, 2004; Mputu et al., 2012a). The necessity of accelerated testing rose, when new long shelf life products where introduced and knowledge about their storage characteristics was required (Mizrahi, 2004). For freeze-dried material the test attempts to accelerate the degradation through elevated temperature (Achour et al., 2001; Morgan and Vesey, 2009) and the cell viability is periodically checked. A product is considered stable when no significant decrease in cell viability during the test time is measured (Morgan and Vesey, 2009). With these results, storage behaviour at lower temperature conditions is then inferred (Achour et al., 2001). Since many of the products are thermodynamically unstable (Franks, 1994) or the tests are made at temperatures close or above glass transition temperature (Achour et al., 2001), the extrapolation can be misleading due to physical and/or chemical changes. The knowledge of possible physical changes during heating allows accelerated testing to be performed with certainty (Franks, 1994), using other models than the Arrhenius one (Achour et al., 2001) to extrapolate data. ASLT could be used to "predict" the storage quality of PPP. Mputu et al (2012) found, that the Arrhenius models deliver results that could predict storage stability or survival rate loss of different *P. fluorescens* strains at various temperatures, however from the text it is not clear if it was from the bacteria with or without CPA added. The authors also found an exponential effect of the storage temperature on the survival rate loss (Mputu et al., 2012a). For formulated *Tsukamurella paurometabola* C-924 the estimation of the shelf life time at 25 °C was calculated by the Arrhenius model from storage test at 25, 32 and 55 °C (Hernández et al., 2006). Also for freeze-dried lactic acid bacteria and probiotics, with ASLT it was possible to predict the survival rate loss of the studied strains during storage and consequently the shelf life (Achour et al., 2001; King et al., 1998).

Influence and optimization of cultural parameters

Variation in the performance of BCA can be due to the low quality or the poor survival in the environment of the microbial product (Palazzini et al., 2016). Information about the influence of culture conditions on quantity and quality of an antagonist is important for commercial production (Daryaei et al., 2016d). The efficacy depends mainly on the strain used, although there may be important physical and nutritional requirements that help the BCA to remain active for a longer time (Keswani et al., 2016). To enhance the biocontrol activity of BCAs, improving them physiologically is a common strategy, since these agents can not adapt to fluctuating environmental conditions when applied in the field (Palazzini et al., 2016). However, most studies on optimization of BCA culture conditions production show the effect on biomass yield regardless of the quality (Daryaei et al., 2016b). The process that produces the highest yield is not always the one that presents micro-organism with the best resistance to the formulation process or the best efficacy (Angeli et al., 2016). In part of this thesis the influence of different fermentation parameters on freeze-dried cells were tested. The growth temperature and time combination showed the highest influence on the efficacy, depending on pathogen strain and assay method (chapter 4). For *T. atroviride* LU132 it was shown that by manipulating the culture conditions, quantity and quality of the produced conidia changed. Conidia production, germination and bioactivity were influenced by the medium composition of nutrient amendments, the carbon content and carbon: nitrogen ratio, temperature, pH and water activity of the medium. The efficacy was tested in dual culture experiments against *R. solani* by measuring inhibition and overgrown activity. The optimum values of the selected parameters differed between conidia production and the ones for the best fitness (Daryaei et al., 2016b,c,d,e). Differences were also found when *T. atroviride* LU132 was tested in greenhouse trials against *R. solani* on ryegrass. The fungus grown at 20 °C in the presence of saccharose in the medium, significantly increased all growth parameters tested compared to the pathogen only application, even though it was not the most rhizosphere competent in comparison to the other fermentation condition (Daryaei et al., 2016a). This shows that manipulation of culture conditions and their interaction can affect the quantity and quality of BCAs (Daryaei et al., 2016d).

Besides the use of protective materials, the type of drying technology and the rate of dehydration, the survival is also affected by the culture medium used for raising the bacteria and the physiological state of the bacteria at harvest time (Bashan et al., 2014; Morgan et al., 2006; Slininger and Shea-Wilbur, 1995). For example, medium and age of cell culture are known to modify cellular freezing resistance, because they determine the values of intracellular glass transition temperature (Fonseca et al., 2016). Generally, it is assumed that cells in the logarithmic phase are less desiccation tolerant than cells within the stationary growth phase. There, the stress response due to carbon starvation and exhaustion of available food sources induces various physiological states within the cells that allow survival of the cell population. However, the optimal growth phase for desiccation survival is also highly micro-organism dependent. There are no generic culture conditions suitable for a range of micro-organisms. The culture conditions have to be optimized at a strain level (Morgan and Vesey, 2009). In storage experiments at 25 °C, the freeze-dried cells of Pfl53 fermented at 28 °C showed a faster reduction in survival after the fifth storage week than the cells cultivated at 20 °C . Survival lower than 1% was reached at week nine for cells fermented at 28 °C compared to the cells fermented at 20 °C where this reduction was shown at week 12. Better survival in time was shown for both temperatures for cells harvested at the beginning stationary phase (chapter 4).

Generally, when producing cells or spores of bacteria or fungi, fermentations studies are done to achieve the maximum production of high quantities of biomass in a short time. Efficacy and storage are then studied in relation to the next step, which means formulation. Fermentations parameters are generally studied when substances produced by micro-organism are of interest like antibiotics (Gao et al., 2016) or antifungal products like phenazine-1-carboxylic acid (He et al., 2008). In fact, to maximize antimicrobial metabolites yield, the optimization of the culture conditions is as crucial (Gao et al., 2016) as for the mass production process (Droby et al., 2016). The optimization of the fermentation medium and parameters, can be done using statistical methods. The Plackett-Burman design (PBD) can be used to identify the factors that have a significant influence on the production of the desired response (Angeli et al., 2016; He et al., 2008; Shu et al., 2017). With the response surface methodology (RSM), with a limited number of experiments evaluation of the effects of several influencing factors and their interaction can be studied (Angeli et al., 2016; Gao et al., 2016; Ghasemi and Ahmadzadeh, 2013; He et al., 2008; Shu et al., 2017). By applying these methods, the maximum dry mycelium weight and conidial yield of *A. quisqualis* ITA3 was assessed. Reduction of powdery mildew symptoms on cucumber and grapevine was also achieved by the conidia produced on the optimised medium (Angeli et al., 2016), however it is not reported if other fermentation condition could still enhance efficacy. With RSM the antibiotics production of *Streptomyces lavendulae* Xjy could be increased by 10% after optimizing the medium constituents, and by 20% after optimization of the fermentation condition (Gao et al., 2016) and consequently also the efficacy. Biomass increase of *B. subtilis* UTB96 and efficacy was achieved applying RSM. However, the efficacy measured as mycelial growth inhibition in dual tests, changed in relation to the antagonist (Ghasemi and Ahmadzadeh, 2013). Additionally, diverse modelling methods give different results as shown for the phenazine-1-carboxylic acid production by *Pseudomonas* sp. M18G (He et al., 2008). RSM was also used to determine the composition of a complex of CPA to ensure the survival of *Lactobacillus rhamnosus* cells (Zhao et al., 2016) and of *Streptococcus thermophiles* (Shu et al., 2017) during freeze-drying. The authors could achieve a significant increase in the survival rate and cell concentration after the process and determine, which of the CPAs and their interaction had the most significant influence (Zhao et al., 2016).

Compatibility studies between biological control agents and pesticides

Growers and extension services are applying a combination of several strategies to consequently shift to less chemicals application (Colla et al., 2012). Knowledge about the impact of chemical pesticides on beneficial micro-organism is an important information to develop a successful IPM program (Castro et al., 2016; EPPO, 2012; Faraji et al., 2016; Vimi et al., 2016) since compatible phytosanitary products are a valid tool for sustainable management (Constantinescu et al., 2014). Interchange of PPP containing micro-organism with chemical pesticides, or other means of protection, could be also a worthy strategy to resistance prevention management (Grabke and Stammler, 2015; Gressel, 2015; Matyjaszczyk, 2015). In general, different diffusion methods are used to test compatibility of micro-organism and antimicrobial products. These are well known in the pharmaceutical industry where they are used for antimicrobial activity screening and evaluating methods (Balouiri et al., 2016). In diffusion methods, the microbial inoculum is spread over a Petri dish and the antimicrobial substance is deposited in or on the agar. Another method is to incorporate the substance that has to be tested into the agar and the micro-organism is inoculated at the centre of the plate (Balouiri et al., 2016). In the first case, the inhibition zone is then measured as reported for different *Bacillus* and *Pseudomonas* strains (Constantinescu et al., 2014; Singh and Dubey, 2010). The second method, also known as poisoned food method (PFM), is mainly used for fungi but examples with bacteria are also reported (Mohiddin and Khan, 2013; Vimi et al., 2016). For fungi, the diameters of fungal growth are periodically measured after incubation under suitable conditions for the strain tested. Beside mycelial growth, sporulation ability can also be tested and compared to the control (Balouiri et al., 2016). With this technique, the inhibitory effects of pesticides on mycelial growth of *Trichoderma* strains (Archana et al., 2012; Kumar and Mane, 2017; Madhavi et al., 2011; Mohiddin and Khan, 2013; Saxena et al., 2014; Singh and Dubey, 2010), *Pochonia chlamydosporia* (Mohiddin and Khan, 2013) and for entomopathogenic fungi like *M. anisopliae* isolates (Bruck, 2009; Faraji et al., 2016) and *Beauveria bassiana* (Faraji et al., 2016) were assessed. For entomopathogenic fungi, the target insect could also be used as bait (Castro et al., 2016; Klingen and Westrum, 2007).

In my studies, bioproducts were tested for their compatibility with chemical pesticides used in strawberry cultivation. RhizoVital® 42 fl., Trichostar® and spores of the *M. brunneum* Ma43 were mixed singly with the chemicals to simulate a hypothetical tank spray (chapter 5). The germination tests followed directly after contact or after four hours. These times were chosen to consider direct spray or transportation time to the field respectively. Constantinescu et al. (2014) for example, mixed *B. bassiana* with chemicals for one hour before germination and vegetative tests. Other authors used a turbidometric method to mimic tank spray application. There, bacteria were shaken in flasks containing broth medium added with the chemical pesticide in different concentrations. The bacteria were considered tolerant when the turbidity, measured by OD, increased in time. With this test method, *P. fluorescens* Pfl with azoxystrobin (Anand et al., 2010; Archana et al.,

2012) and *P. aeruginosa* isolate GSE18 and isolate GSE19 with chlorothalonil were tested (Kishore et al., 2005). Also the growth kinetic of *B. cereus* strain JY9 and *Methylobacterium* sp. HJM27 in the presence of organophosphates and biopesticides was assessed (Sultan et al., 2013). With this test method, it was shown that biocontrol bacteria like *P. fluorescens* and *Bacillus* spp. are tolerant to higher concentrations of chemicals and can use pesticides as nutrients (Kumar and Mane, 2017; Mohiddin and Khan, 2013; Myresiotis et al., 2012).

As seen there is no universal standard test which makes a comparisons difficult. At any rate, all methods showed just mimicking of the environment where the bioproducts will be applied. Bruck (2009) found that when the fungicides were applied with *M. anisopliae* (F52) spores incorporated into soilless potting media, the impact was less than in *in vitro* tests. Positive interaction, meaning stabilization or increase in CFU/ml in time, was also shown for different *Bacillus* strains grown in the presence of pesticides in sterile soil trials (Myresiotis et al., 2012). Castro et al. (2016) stated, that if a pesticide is considered to be compatible in laboratory bioassays at maximum exposure, it can be assumed that these pesticides result in the lowest impact on beneficial micro-organism. In contrast, if a pesticide has negative effects in laboratory bioassays, it may not always necessarily result in negative effects in the field. This occurs because the fungi can hide within the crop (Castro et al., 2016), debries or in the target organism (Knudsen and Dandurand, 2014) and a pesticide applications does not necessarily result in full coverage (Castro et al., 2016). On the other hand, pesticides may reduce microbial competition with an increase in nutrient availability for BCA as shown for *P. aeruginosa* isolates, applied in the field with low doses of chlorothalonil (Kishore et al., 2005). Consequently, if compatibility is shown, pesticide residues in the soil will not affect the efficacy of BCAs. This implements the possibility of integration between micro-organisms and chemical pesticides for the control of soil-borne plant pathogens (Kumar and Mane, 2017; Mohiddin and Khan, 2013). On the other hand, if a growth reduction is shown, reducing the chemical pesticides in the agro-ecosystem would decrease biotic stress, maintaining a natural ecological balance and eventually better crop yield (Sultan et al., 2013). For *P. aeruginosa* isolates GSE18 and GSE19, for example, their combination with chemicals was more effective at control of late leaf spot and increased yield than each treatment alone (Kishore et al., 2005). In any case, the reported statements have to be applied with caution until more exhaustive field studies are conducted (Castro et al., 2016). In my studies, the commercial field trials were tilled by the farmers according to their individual techniques and pesticide schedules. I did not test the micro-organism population's behaviour over time of the applied products but, when yield is considered, the results seem to show no massive influence of the chemical pesticides on micro-organisms since, except for one application, the yield increased. It can be assumed that the environment has more influence on the BCA than chemicals, except for some fungicides that could influence fungal BCAs (chapter 5).

From the screening to the applicable microbial plant protection product

Screening methods are selective and only a portion of the possible antagonists will be detected. Generally, the preliminary screening, e.g. dual cultures, are the least laborious and most artificial, while the most natural and laborious, e.g. in the field, are the finals. Therefore, it can happen that the BCA candidate with better efficacy under field conditions can be missed (Knudsen et al., 1997). Consequently, the objectives of the researcher influences the screening process (Pliego et al., 2011).

In vitro *dual culture and on detached plant parts tests*

Assays to assess the potential of BCA against the target pathogen are a key point during PPP development. Screening for antagonistic ability is generally made *in vitro*. In dual cultures, the potential BCA isolates and the pathogen are streaked, or a disc of actively growing mycelia is put at one side of a Petri dish. After a given time the inhibition zone and/or mycelial growth is recorded and the percentage of inhibition can be calculated (Maurya et al., 2014). Only BCAs with a positive inhibition are then carried forward (Comby et al., 2017; Shehata et al., 2016) so that potential antagonists could be dismissed inadvertently (Pliego et al., 2011). On the other hand, these screenings can identify micro-organisms as antagonist that later in *ad planta* trials show no effect (Daayf et al., 2003; Shehata et al., 2016). In general, dual culture screenings detect micro-organisms that secrete substances that interfere in the targeted pathogen's life cycle, like antibiotics and enzymes (Comby et al., 2017; Pliego et al., 2011), or act per direct antagonism or competition for agar nutrients (Shehata et al., 2016). Antagonists that act by inducing host resistance or compete for ecological plant niches are not detected (Pliego et al., 2011; Shehata et al., 2016). Since quantification of inhibition zones is used as an indicator of competitive ability of BCA and some species could inhibit fungal activity without creating them, Cray et al. (2015) developed an inhibition coefficient that considers the contribution of different inhibition types on fungal growth rate. The resulting quantified antifungal activity of the considered BCA, was consistent with their known potency (Cray et al.,

2015). Other methods evaluate the "production of compounds related to biocontrol and/or promoting growth" and take them into account to reduce the number of evaluations (Berg et al., 2001; Mota et al., 2017; Shen et al., 2016). According to the target pathogen, to know which compounds are produced, can help to select the best candidates. In some cases, also the application of particular screening strategies can fail the selection of the right BCAs (Pliego et al., 2011). Another limit of this commonly used *in vitro* screening test is shown when the level of antagonism depends on the growth rate of the strain. This should be considered when ranking the potential antagonists by dual culture assays (Comby et al., 2017). For example, when *Pseudomonas* sp. strain PsJN and *B. cinerea* were simultaneous inoculated, the bacterium did not stop the growth of the fungus but with two days' advance, a clear inhibition appeared. This could be caused by toxin production but also because *Pseudomonas* grows slower than *B. cinerea* (Barka et al., 2002).

For the studies presented here, dual culture was one of the selection criteria for the successive studies in greenhouse and field trials. Ninety-six potential antagonists were tested against four soil-borne pathogens (*V. dahliae, V. albo atrum, P. cactorum* and *P. fragariae*) known to infect strawberries and to be a problem in Germany. In Europe, *Verticillium* and *Phytophthora* complexes are considered an important target for biocontrol products since actually no biological solutions are under development. Screening of new BCAs is considered one priority (Lamichhane et al., 2017), however tests with commercial BCAs already registered are also an option (Robinson-Boyer et al., 2009). The presence on the market as PPP or strengthener was another selection criterion for the micro-organism used in the experiments on strawberries, reported in this thesis. Dual culture was also used to assess the influence of the production parameters temperature and time on Pf153 efficacy against different strains of *B. cinerea*. The freeze-dried cells of Pf153 were spread on the surface of the whole agar plate and a plug of actively growing pathogen was directly deposited into the middle. Mycelial growth was measured on selected days (chapter 4). Since *B. cinerea* infects above soil parts and an application of BCA would be done by spraying, spreading the antagonist on the plate was chosen to partially mimic real conditions. The tests were made on two media containing different sugars allowing investigation of different nutrient conditions. One medium had a composition comparable to strawberry flower extracts (Hjeljord and Strømeg, 2004) and was used to simulate a more natural environment. Indeed, on different culture media, antibiotic production can vary and one culture medium may be not enough to identify some antibiosis activity (Daayf et al., 2003). For *B. cinerea*, different behaviour between strains in the presence of different fermented Pfl53 was seen, but these were not statistically significant. The tests were performed with formulated Pfl53. Other authors have tested micro-organisms contained in products known to be antagonistic to other diseases however after re-isolation of the micro-organism (Khan et al., 2011). The influence of CPA on performance is described in the forgoing section "Co-formulation substances".

One step forward to real application, and a quick confirmation method for the activity previously shown on Petri plates, is to use detached leaves. However, detached leaves have limited resources and the system has still limitations when induced resistance is the main mode of action (Daayf et al., 2003). In my own experiments detached *V. faba* leaves were used where Pfl53 and *B. cinerea* were sprayed on the leaves. Spraying the pathogen directly after antagonist application, without waiting for a bacterial establishment on the plate or leaf was done to mimic natural behaviour, like using no disinfected leaves. Following the same method, the biocontrol efficacy of the two chemical products against *B. cinerea* on strawberry detached leaves was demonstrated (Robinson-Boyer et al., 2009). Detached leaf assays are reported for diverse plants/pathogens combinations differing in the inoculation procedure. Tests against *B. cinerea* were performed on surface-sterilized detached leaflets of tomatoes, where bacterial suspension was sprayed on the leaves but the pathogen was inoculated by needle-prick wounds (Kefi et al., 2015). For strawberries, leave discs were used (Helbig, 2002) but often the tests were made with flowers, since infection during blooming is often the origin of gray mould on fruits. The collected strawberry flowers were sprayed to runoff with the antagonist, afterwards inoculated with droplets containing *B. cinerea* conidia (Helbig, 2002; Hjeljord et al., 2001). In tests with *Pseudomonas* against *Pyrenophora teres*, drops of the micro-organism suspensions were used to inoculate detached barley leaves (Khan et al., 2010). Potato leaves and wheat spikelets were dipped in antagonistic suspension prior to inoculation with drops of *Phytophthora infestans* (Daayf et al., 2003) or spray of *Fusarium graminearum* conidia (Comby et al., 2017) respectively. Afterwards, the plant parts were put into a sealed container at suitable growing conditions for the pathogen. A visual assessment of the disease is generally made after an appropriate time, depending on the growth velocity of the pathogen (Comby et al., 2017) as made in my experiments, or by measuring the diseased area or pathogen growth as for dual culture tests.

Experimental designs for greenhouse and field trials

Generally, greenhouse experiments follow laboratory efficacy tests before field trials. In the field, due to variable abiotic and biotic parameters, the experiments may be difficult to evaluate (Knudsen et al., 1997). Experiments in a closed and controlled environment allow the study of a complex system, reducing the number of parameters that can influence the activity of added micro-organisms. Considering contemporary external and internal influences makes the interpretation of the trials to complex since too many factors can influence the BCA/plant/pathogen constellation. On the other hand, research restricted to one or two parameters can lead to false results as shown for the ability to produce secondary metabolites on Petri plates or on plants by antagonistic strains that is likely to differ (Comby et al., 2017). Tests in pots, where field conditions are simulated by using natural soil, can be considered a compromise (Knudsen et al., 1997). Greenhouse experiments are useful to obtain information about the mode of action as antagonism (Pliego et al., 2011), because tests based on whole plants have a higher number of components, are closer to natural conditions and more realistic than *in vitro* or detaches leaves tests (Daayf et al., 2003; Pliego et al., 2011). However *in vivo* selection is not simple due to the different interactions between BCA and plant (Mota et al., 2017). Failure in disease suppression by BCA candidates between *in vitro* and *ad planta* can be due to failure in plant colonization, migration to the pathogen, competition with native microbiota and/or differences in production of anti-fungal compounds between the tests systems (Shehata et al., 2016). For example, antagonistic *Gliocladium* isolates found in preliminary screening under controlled environmental conditions, did not have the same biocontrol performance in the field (Knudsen et al., 1997).

For my experiments, the micro-organisms selected by dual culture mentioned above were used in greenhouse and field trials with strawberry plants (chapter 5). The strawberry cultivar Honeoye was chosen because it is very sensitive to *V. dahliae* (Harris and Yang, 1996; Olbricht and Hanke, 2008), susceptible to *P. cactorum* (Anandhakumar and Zeller, 2008; Eikemo et al., 2000) and among early ripening cultivars adapt to organic production due to high yields (Daugaard and Lindhard, 2000), making it the perfect candidate for the experiments. The pathogen inoculation was made by mixing the pathogen growing media containing the propagules into the potting soil for greenhouse trials, or adding it to the planting hole for field trials. A mixture of four *V. dahliae* strains, re-isolated from strawberries plants with different grades of aggressiveness, were grown in separate Erlenmeyer flasks and mixed before use. Mixing pathogen substrate with the potting soil was also used to inoculate *R. solani* in greenhouse trials with ryegrass (Daryaei et al., 2016a) and strawberries (Elsaid et al., 2005). Also *V. dahliae* (Berg et al., 2001), *P. cactorum* and *P. fragariae* propagules were added in this way in strawberry experiments (Anandhakumar and Zeller, 2008). In other cases, the infection was achieved by adding a suspension of the pathogen as shown for *M. phaseolina* and *F. solani* where the pathogens were added at the base of the pot where strawberry were planted (Pastrana et al., 2016). Root dipping or soil drenched with a conidial suspension is often used for *Verticillium* infection in tests using sterile soils (Deketelaere et al., 2017) as shown in greenhouse and field trials with potatoes (Uppal et al., 2008) and strawberries (Diehl et al., 2013; Kuchta et al., 2008). Naturally infested field conditions however, are quite different from these experimental conditions (Deketelaere et al., 2017) since the natural infection is made by encountering of the root with the *Verticillium* MS. In fact, for *Verticillium*, it is reported that naturally infested soil or MS should be used for trials in field and greenhouse conditions (Deketelaere et al., 2017) as done in the trials presented here. On the other hand, root dipping in a conidial suspension showed to be more consistent than mixing MS in the soil (Gordon et al., 2005). To mix different strains of *V. dahliae* for the purpose of assuring the effectiveness of inoculation was also reported in potato (Uppal et al., 2008) and in mint trials for *Verticillium longisporum* (Depotter et al., 2017). In my trials, the *V. dahliae* mixture did not show a particularly high infection grade and no plants died during the greenhouse experiment. The growing parameter tested did not show differences between pathogen inoculated and non inoculated plants, but root dry weight and the number of daughter plants were increased. Reduction in yield was observed. On the other hand, for *P. cactorum* a single strain was used which showed high aggressiveness. At the end of the greenhouse trials, 44% of the plants in the control were dead and its presence reduced most of the growing parameter assessed (chapter 5). Reduced aggressiveness of the *V. dahliae* mix can be due to different causes. For BCA, it is reported that the co-inoculation with two or more strains does not necessarily bring an additive or synergic effect because through competition, the expected positive influence obtained by single strain inoculation can be reduced or even disappear (Pliego et al., 2011; Robinson-Boyer et al., 2009; Trabelsi and Mhamdi, 2013; Vestberg et al., 2004; Yang et al., 2014). Also the introduction in the environment can affect the performance in the rhizosphere by competition with the indigenous microflora (Yang et al., 2014). This could be an explanation for the poor results obtained by the co-inoculation of the four *V. dahliae* strains, competition between strains could have reduced their infection capacity.

In my experiments, the antagonists application was made by root dipping. Plants were dipped, prior to planting, for a given time in the microbial suspensions, as it is done for the chemically active substance Fosetil-Al used against *Phytophthora* sp. in strawberry growing. This system was also used by other authors in field trials (Anandhakumar and Zeller, 2008; Diehl et al., 2013; Kurze et al., 2001; Pastrana et al., 2016; Porras et al., 2007). Antagonist application can also be done by watering the plants with the microbial product suspension directly after planting (Pastrana et al., 2016). The advantage of root dipping is that the roots are directly exposed, free from soil, to the antagonistic inoculum (Anandhakumar and Zeller, 2008). Application of the antagonist close to the roots could be an effective strategy where the disease initially infects the plant roots as Verticillium wilt (Deketelaere et al., 2017).

The substrate used in greenhouse trials is generally pasteurized or sterilized before use (Berg et al., 2001; Daryaei et al., 2016a; Knudsen et al., 1997; Palazzini et al., 2009; Pastrana et al., 2016). In sterile soils, BCA establishment is easier than in the field because the conditions there are less complex. This could be a reason why BCAs often fail to work afterwards in real conditions. On the other hand, disease develops fast and often with more severe symptoms, so that some effective BCAs are rejected because they seem of minor importance during selection in sterile conditions (Deketelaere et al., 2017). In my greenhouse trials, the soil used was not pasteurized before mixing it with the pathogen. Temperature and light were monitored. In these experiments, the applied antagonists gave fluctuating results for the values of the selected growth parameters, alone and in the presence of the pathogen, and none showed superior results overall (appendix to supplement 4). In my controlled field trials, the performance of the antagonists was also poor considering the strawberry yield, which was in many cases reduced (appendix to supplement 4). Evidence that yield and quality of the harvested product is not negatively affected is needed to support the registration (EPPO, 2012). These poor results could be caused by the addition to the field of compost soil before starting the trial, which could have reduced the biocontrol efficacy. High levels of microbial biomass in soil can increase the interaction between BCA and soil micro-organisms and reduce their efficacy (Bae and Knudsen, 2005). Other authors also showed no positive effect of added BCA. An increase in the number of dead plants compared to control samples and a non homogeneous effect was shown for a mixture of non-pathogenic strains of *V. dahliae* used as BCA in field trials

(Diehl et al., 2013). Lower yields after application of *T. harzianum* P1 were also found in strawberry trials against *B. cinerea*. There, the antagonistic fungus was applied on flowers at different stages or on green berries without any effect on the percentage of healthy fruits (Hjeljord et al., 2001). To increase the efficacy, an activation of *T. harzianum* P1 was tested, however no significant difference was found between quiescent and activated conidia (Hjeljord et al., 2001). Also in my trials, Trichostar® was activated by suspending the product in lukewarm tap water for four hours before application, as recommended by the producer. The liquid formulation contains saccharose. It was shown that *Trichoderma* grown in saccharose at 20 °C significantly increased the growth parameters of ryegrass plants in greenhouse trials. Applied in the presence of *R. solani* it suppressed the disease, especially at lower pathogen concentration (Daryaei et al., 2016a). Since germination of *Trichoderma* isolates requires exogenous nutrients, it is assumed that the ability to exert competitive biological control is enhanced by their addition. Unclear is, if nutrient availability affects other antagonistic mechanisms such as antibiosis and mycoparasitism or the expression of some cell wall-degrading enzymes. Therefore, the interactions between *T. harzianum* and pathogens that germinate rapidly is a topic not to be overlooked (Hjeljord et al., 2001).

Importance of testing the formulated micro-organism under real commercial conditions

For the success of BCA, an acceptable and consistent performance under commercial conditions is crucial. In order to compete with conventional methods (Gullino, 2005), microbial biopesticides should have an efficacy comparable to chemical fungicides (Droby et al., 2016; Lamichhane et al., 2017), be effective under unfavourable environmental conditions and provide stable and consistent results (Gullino, 2005). Research to answer the practical problems posed by the growers is important and therefore the strategies proposed should be validated under different situations like in field studies and on-farm validation (Colla et al., 2012; Gamliel, 2010). This is also a requirement reported in the regulation 1107/2009 where the efficacy of a PPP has to be shown in applications consistent with good practice and in realistic conditions (European Commission, 2009b). Dual tests are generally fast, easy, with a high throughput and allow to test many different antagonists, but the results are not directly reportable to field performance (Daayf et al., 2003; Pliego et al., 2011; Shehata et al., 2016). Under laboratory or controlled conditions, BCAs have shown many times to be effective in controlling plant diseases but then they fail under practical conditions (Teixidó et al., 2006). For example, the physiological improvement of *B. subtilis* RC 218 that increased biocontrol effectiveness in greenhouse experiments, was less effective under field conditions (Palazzini et al., 2016). In fact, reduced biocontrol efficacy in the field could be due to abiotic and/or biotic factors that are more difficult to regulate than in an environmentally controlled greenhouse (Uppal et al., 2008). For example, the bacterial isolate M1 was effective against *V. dahliae* on potatoes for two consecutive years in greenhouse trials but the results under field conditions performed at different locations during two growing seasons, were inconsistent (Uppal et al., 2008). Biocontrol efficacy in natural soils is variable (Knudsen and Dandurand, 2014; Robinson-Boyer et al., 2009) and erratic results in commercial fields are often reported. Environmental conditions, which can be less favourable to the establishment and activity of the added BCA, can be a cause of this variability (Ambrosini et al., 2016; Ji et al., 2006; Lamichhane et al., 2017). Therefore studies in homogeneous environments may be less applicable to predict growth and behaviour of BCAs in a heterogeneous environment (Knudsen and Dandurand, 2014; Uppal et al., 2008). Interactions between pathogens and antagonists are influenced by a huge number of factors, so that it would be necessary to assess them on a case by case basis (Dickie and Bell, 1995). For example, most of the secondary metabolites produced in the rhizosphere environment by plant pathogens are probably unknown, because the compounds produced there tend to differ from the ones produced in isolated cultures like in laboratory tests (de Souza et al., 2016). Also to consider is that metabolites, produced by bacterial cells, can be used by fungi and this can lead to a plant pathogen infections promoted by BCA with eventually reduction of crop yields. This indicates that depending on the prevailing conditions the biocontrol activity is more a mode of behaviour rather than an inherent property of a bacterial strain (Bardin et al., 2015; Cray et al., 2016). The interactions between microbial cells in natural habitats are dynamic. The results of these interactions can vary with infinite outcome combinations (Cray et al., 2015). That is why results between *in vitro* and *in vivo* antagonistic assays can be contradictory (Dickie and Bell, 1995) as shown for the antimicrobial activity against *Verticillium* made *in vitro*, which correlates poorly or not at all with biocontrol activity tested *in planta* (Deketelaere et al., 2017). On the other hand, some authors found that the antagonistic effect *in vitro* was also obtained in controlled environment and/or field conditions (Kefi et al., 2015; Pastrana et al., 2016; Shehata et al., 2016). Tests that use whole plants are more realistic and would be more appropriate for BCAs screening and for mode of action studies (Daayf et al., 2003).

Conditions that mimic the field situation, are expected to significantly improve the chance of successful application in practice (Deketelaere et al., 2017). Therefore, the biocontrol activity of candidate antagonistic strains has to be tested over a full range of field relevant parameters (Cray et al., 2016) and it is important to test their effectiveness when applied in natural environments (de Souza et al., 2016; Pliego et al., 2011). To study the potential of BCA under variable environmental conditions, multi-location field trials across various crop seasons, seems to be unavoidable (Khan et al., 2010; Yang et al., 2014), since their narrow range of activity already limits their commercial success (Droby et al., 2016; Teixidó et al., 2006). In the trials reported in this thesis, the biological products were tested under real conditions. The influence of the two plant strengthener RhizoVital® 42 fl. and Trichostar® was assessed parallel in two fields in two consecutive growing seasons, because in Germany, strawberries are mainly grown as two years crops, in commercial fields (BZFE, 2017). The selected micro-organisms were tested in fields naturally infected with strawberry diseases (chapter 5). Applied by root dipping, the successive treatments were made by irrigating with a watering can which allowed to treat each row independently without cross contamination. This technique simulates a direct application through a watering system. Soil application is generally used as curative measure (Agustí et al., 2011; Pastrana et al., 2016).

Too high residues of the inoculum (Helbig, 2002; Uppal et al., 2008) paired with extremely favourable weather conditions for disease development (Anandhakumar and Zeller, 2008) may be an explanation of the poor performance of some treatments under field conditions. Also the interactions between cultivar/BCA and cultivar/pathogen (Uppal et al., 2008), root colonization competition with the soil microbiome may be responsible for the variable performance (Anandhakumar and Zeller, 2008). The inconsistency of BCA can also be a result of the variable sensitivity of plant pathogens to them, like for chemicals. If plant pathogens have the ability to produce natural mutants through the selection pressure present by the products used in the field, a build-up of field resistance could arise (Ajouz et al., 2011; Bardin et al., 2015). In the field the infection may involve different strains (Ji et al., 2006) of the pathogen and the disease control can be limited by the virulence of naturally occurring strains (Anandhakumar and Zeller, 2008). Differences in the protection level may be provided by BCAs (Ji et al., 2006) and low susceptible pathogen isolates can be selected in the natural population as a consequence (Ajouz et al., 2011). Therefore, it is important to test BCA against different pathogen strains.

As seen, a micro-organism functioning under laboratory conditions, is not always able to produce equivalent results under field conditions, particularly when the formulation was not considered (Nehra and Choudhary, 2015; Uppal et al., 2008). In fact, lack of efficacy of microbial products can be due to wrong storage, transport conditions, improper application (Matyjaszczyk, 2015) but also formulation. In older reports about field applications, the trials were made with fresh fermented cells (Anandhakumar and Zeller, 2008; Khan et al., 2010; Kurze et al., 2001; Uppal et al., 2008), fungal suspensions (Hjeljord et al., 2001) or culture supernatants. The micro-organisms tested in this thesis were used as formulated products and upon delivery they were stored and used as indicated by the producer. As described in section "Co-formulation substances", the formulation influences the efficacy of the BCA and the awareness about the importance to tests the commercial product has increased. Lately the publication of reports where the tests in field conditions are made with formulated micro-organism are increasing showing the importance of considering the end product. For example, the performance of *B. amyloliquefaciens* QL-18 to control tomato wilt in field trials was significantly increased when amended with rapeseed cake and changed due to the carrier used (Wei et al., 2015). Spray-dried cells of *B. subtilis* RC 218 and *Brevibacillus* sp. RC 263 were reported to reduce incidence and disease of *F. graminearum* in semi-controlled field trials on wheat (Palazzini et al., 2016). Increase in strawberry plant fresh weight in greenhouse trials and fruit yield in field trials was achieved by addition of *S. hygroscopicus* B04, fermented on bioorganic fertilizer, to soil seriously infested with *F. oxysporum*. Additionally, the disease incidence was reduced, the rhizosphere microbial community structure altered and the fungal populations in rhizosphere soil significantly reduced (Shen et al., 2016). *P. putida* Rs-198 formulated as liquid bacterial fertilizer significantly improved germination rate of cotton seedlings and increased cotton growth compared with the control plants, where no bacteria was added (He et al., 2015). Also the type of application form of the end product influences the microbial activity as shown for *T. harzianum* (Elsaid et al., 2005) and *Serratia plymuthica* HRO-C48 (Müller and Berg, 2008).

Technologies to understand and improve plant protection products performance

To achieve practical success of biopesticides, understanding the formulation process is as important as the study of physical, chemical and biological environments. As seen, the complex ecology where BCAs are released, influences the field performance of the formulated product (Singh and Arora, 2016) and success in biocontrol of plant diseases in field crops has been limited, due to the complex ecological processes involved (Wei et al., 2016). In fact, under commercial conditions the plant-soil interaction remains a black-box when regulation measures are applied (Diehl et al., 2013). The actual impact of microbial inoculation on agricultural systems remains unknown, varying according to geographical location, plant species and micro-organisms used (Ambrosini et al., 2016). To understand the ecological characteristics of BCAs, is profitable for the development of strategies to maximise the efficacy in field conditions (Wei et al., 2016).

The study of the biocontrol properties is expedited by genome availability that allows high-throughput analyses (Massart et al., 2015b). With omics technologies, time and energy efforts can be reduced during selection and evaluation of potential BCA (Pliego et al., 2011). By characterizing genes, mRNAs and proteins, the modes of action of a BCA can be identified, since molecular approaches allow in-depth characterization of the strain. The combination with traditional microbiological methods improves the interpretation of the genome and the characterized genes of the potential BCAs (Massart et al., 2015b). Culture independent approaches, to find functionally interesting micro-organisms, can be made by using genetic markers and molecular profiling (Pliego et al., 2011). Transcriptomic are useful to understand how BCA act, their gene regulation sustaining the effect on pathogenesis, the plant defence pathways. This method applied to thritrofic level studies, would be useful to understand *in situ* the complex interplay between two or three species. *In vitro* experiments can produce bias (Massart et al., 2015b), the performance of a BCA results from a complex mutual interactions between all system components. The level of expression of the genes that regulate biocontrol genes can change during mass production, formulation and storage which could now be followed due to the development in proteomics and functional genomics (Droby et al., 2016).

A newly emerging field in agriculture is nanotechnology where pesticides or fertilizers, are encapsulated for their controlled release in soil (Singh and Arora, 2016). Nanotechnology would be an option to produce new BCA formulations with superior stability and efficacy (Keswani et al., 2016; Singh and Arora, 2016). Nanoformulations can protect BCA from adverse environmental conditions, through the development of highly stable, effective and eco-friendly pest management practices (Singh and Arora, 2016).

To improve the efficacy of BCAs, diverse possibilities are given. Beside production and formulation, manipulation of BCAs or the addition of a 'helper' strain that enhances activity can improve the antagonistic efficacy (Berninger et al., 2018). Since most of the demanded characteristics for an optimal biocontrol are not present in one organism, hybridization allows the combination of positive traits of different strains or species. Very often, strains with biocontrol efficacy do not have good shelf life or the function parameters are different to those of the pathogen. By crossing strains expressing the diverse desirable attributes a progeny with the combination of desirable traits can be developed (Jensen et al., 2016). Genetic modification of beneficial microbials would be also a valid approach (Jensen et al., 2016) but regulatory constraints and public concerns are hurdles to the use of genetically modified organisms (Droby et al., 2016), as seen for transgenic plants (Singh and Dubey, 2017). Using BCAs as a useful genes source may have a potential for creating genetically modified plants (Jensen et al., 2016).

Verticillium wilt and microsclerotia quantification

Verticillium wilt is becoming a more relevant problem in different crops (Deketelaere et al., 2017). It is difficult to control since there are no curative fungicides available (Depotter et al., 2017). MS are its resting structures, that after germination reach the root surface with their hyphae and enter the young roots with a consequent systemic infection. The blockage of the vascular system through dispersion of *V. dahliae* conidia and mycelial growth inhibits the flow of water and nutrients. MS are formed near colonized vascular tissues in a few weeks upon death of the host plant (Eynck et al., 2007; Mol et al., 1996; Pascual et al., 2010). Since MS are the primary source of *Verticillium* disease in the field, it is recommended to quantify the present MS before planting. MS are distributed in clusters and this has to be considered when the field has to be probed (Neubauer and Heitmann, 2011). In one field used for the trials, the cluster presence was shown when we probed six successively planted strawberries. The first three and the fifth were heavily diseased and the fourth and sixth were still vigorous and did not show disease symptoms. Therefore, it is necessary to find a balance between number of probes that are representative for the tested area, the accuracy of the results and the time and effort invested (Neubauer and Heitmann, 2011). In general, the results obtained are a mean value and no spatial distribution information is given. For the results presented here, four probes for each replicate of each treatment, 16 probes for 24 m or 100 strawberry plants, were taken and mixed. The detection of the MS was made by the wet-sieving method described by Neubauer and Heitmann (2011) which is also used by the agricultural authorities to give recommendations.

To quantify MS in soil, different methods are described. These methods are highly soil-type dependent and large differences between laboratories in the quantification can be observed even when the same method is used (Goud and Termorshuizen, 2003). The methods available are plating methods (used in risk analysis studies), bioassays, molecular methods and immunoassays. Plating methods are divided in dry (Short et al., 2015) and wet (Steffek et al., 2006) and both involve spreading of soil onto a Petri dish containing semi-selective agar media. After incubation, the new MS developed colonies are counted under a dissecting microscope (Goud and Termorshuizen, 2003). This determination method can take several weeks to be completed (Bilodeau et al., 2012; Debode et al., 2011; Gharbi et al., 2016) and to differentiate between *Verticillium* species that produce MS by microscopic analysis is not easy (Debode et al., 2011; Gharbi et al., 2016). New PCR-based detection methods can significantly reduce the processing time (Chellemi et al., 2016; Debode et al., 2011), can differentiate among the MS forming *Verticillium* using specific primers (Debode et al., 2011) and offer potential to reduce costs (Chellemi et al., 2016). In particular, for a multiplexed TaqMan real-time PCR assay (Bilodeau et al., 2012) and a density flotation-based extraction of MS followed by real-time PCR assays (Debode et al., 2011) reproducibility and sensitivity were shown for a range of field soils. A newly developed Q-PCR assay was reliable in detecting significant differences between severity stages, with no influence of the nonviable MS on the final quantification result (Gharbi et al., 2016). In addition, since *V. dahliae* has also a host preference in relation to its vegetative compatibility of individual isolates, with PCR the predominant virulence type could also be determined (Zeise and von Tiedemann, 2002). As for other methods, some factors can interfere in the PCR based detection of fungus deoxyribonucleic acid, for example plant components like polysaccharides (Kuchta et al., 2008).

The quantification of MS density in soil is important for disease prediction and for assessing the effect of control measures. Several plant hosts of *V. dahliae* show a direct relationship between the inoculum density in soil and progression of the disease (Gharbi et al., 2016). The relation between measured inoculum densities and observed disease or yield loss of the crop, is affected by soil microbial, chemical and physical factors. Therefore, a quantification method which is affected by the same factors could better predict the disease infection than a method that retrieves 100% of the MS (Goud and Termorshuizen, 2003). In addition to the quantitative assessment, the determination of the predominant virulence type can help to estimate the danger deriving from *Verticillium* soil infestation (Zeise and von Tiedemann, 2002).

Commercially acceptable strawberry varieties, exhibit limited levels of tolerance to Verticillium wilt in the field, which limits host resistance as an effective tool for disease management (Chellemi et al., 2016). A successful BCA against *V. dahliae* should reduce MS and the wilt incidence, share the same ecological niche, induce resistance responses in the plant, promote plant growth and improve yields of infected plants close to healthy plants or crops grown in disease free soil (Deketelaere et al., 2017; Klosterman et al., 2009). BCAs often promote shoot and/or root growth. The plant growth promoting effect can compensate the deleterious effect of pathogenic *Verticillium* species on the yield of crops (Deketelaere et al., 2017).

For the concentrations of MS that we found at the beginning of the trials in the commercial fields (7.2 and 8.8 MS/g soil) the infection risk is considered high (Neubauer and Heitmann, 2011). Through laboratory analysis, the number of MS was followed during the two testing years. The microbial treatments could constantly reduce the number of MS in one field but not in the second field. In some cases the application of the microbials increased the number of MS, however a general positive increase in strawberry yield was found (chapter 5). Fluctuation of MS numbers were also found when allyl-isothiocyanate was added by drip irrigation to terminate an infected strawberry crop and after incorporation of crab/feather meal before planting broccoli. In that case the oscillations were associated with the redistribution of MS in the soil profile due to tillage practice (Chellemi et al., 2016). Similar variable results were found for *P. cactorum* oospores. In this case, it was supposed that the differences in the effect of the added micro-organisms could be caused by the physiological status of the micro-propagated strawberries (Vestberg et al., 2004). Since field diagnosis by visual rating is difficult (Koike and Gordon, 2015), in my trials, affected plants were evaluated without categorizing the disease. The number of symptomless plants in the commercial fields was generally higher than the control, as the number of dead plants. In some cases, the yield of the plants treated with the bioproducts was reduced when compared to the control. Since the cv. Honeoye shows some yield fluctuations from year to year (Daugaard and Lindhard, 2000), this could be a partly cause for the reduced yield. Some cases of yield reduction or equal yield to the control was also reported when *T. asperellum* and *Bacillus* spp. formulation were applied against *M. phaseolina* and *F. solani* in strawberry field trials (Pastrana et al., 2016).

CONCLUSIONS

There are many scientific reports about potential BCA but it is still difficult to find new biological PPP on the market. Micro-organisms are living organisms that are influenced by many biotic and abiotic factors, from the fermentation to the application. The production and formulation processes, in particular for non-sporulating bacteria like *Pseudomonas fluorescens*, are still a challenge and have to be optimized for each strain. The resulting product is a compromise between biomass production, storage and efficacy capabilities. Many BCA that show antagonism in laboratory or controlled chambers can fail their ultimate purpose, effectiveness in the field, because during efficacy tests they are not used in their end formulation and in their application environment.

The results presented here show that not only the formulation and conservation modalities but also the fermentation parameters influence the storability and the efficacy of Pf153. Differences in the performance where also found regarding the testing system confirming the importance of efficacy trials with the end product in the end application environment. Selected micro-organism also found as commercial bioproducts, showed a good performance in laboratory dual tests but failed in greenhouse and field trials where the soil-borne pathogens *Phytophthora cactorum* and *Verticillium dahliae* were artificial inoculated. In commercial fields infested by soil-borne pathogens, an increase in strawberry yield of about 8% for Trichostar[®] and 6% for RhizoVital[®] 42 fl. was shown. The bioproducts were effective in two different fields in two consecutive years, showing that before application, it is important to consider soil properties and field instrumentation. Also the compatibility with diverse chemical pesticides was shown.

Biological PPP can be an important tool in IPM. Through collaboration between industry, science and farmers the effort could bring a product with stable results in time and different environment.

REFERENCES

Abadias, M., Benabarre, A., Teixidó, N., Usall, J., and Viñas, I. (2001). Effect of freeze drying and protectants on viability of the biocontrol yeast *Candida sake*. *International Journal of Food Microbiology* **65**(3):173-182.

Achour, M., Mtimet, N., Cornelius, C., Zgouli, S., Mahjoub, A., Thonart, P., and Hamdi, M. (2001). Application of the accelerated shelf life testing method (ASLT) to study the survival rates of freeze-dried *Lactococcus* starter cultures. *Journal of Chemical Technology and Biotechnology* **76**(6):624-628.

Agustí, L., Bonaterra, A., Moragrega, C., Camps, J., and Montesinos, E. (2011). Biocontrol of root rot of strawberry caused by *Phytophthora cactorum* with a combination of two *Pseudomonas fluorescens* strains. *Journal of Plant Pathology* **93**(2):363-372.

Ajouz, S., Walker, A. S., Fabre, F., Leroux, P., Nicot, P., and Bardin, M. (2011). Variability of *Botrytis cinerea* sensitivity to pyrrolnitrin, an antibiotic produced by biological control agents. *BioControl* **56**(3):353-363.

Akhtar, M. S., and Azam, T. (2014). Effects of PGPR and antagonistic fungi on the growth, enzyme activity and Fusarium root-rot of pea. *Archives of Phytopathology and Plant Protection* **47**(2):138-148.

Ambrosini, A., de Souza, R., and Passaglia, L. M. P. (2016). Ecological role of bacterial inoculants and their potential impact on soil microbial diversity. *Plant and Soil* **400**(1-2):193-207.

Anand, T., Chandrasekaran, A., Kuttalam, S., Senthilraja, G., and Samiyappan, R. (2010). Integrated control of fruit rot and powdery mildew of chilli using the biocontrol agent *Pseudomonas fluorescens* and a chemical fungicide. *Biological Control* **52**(1):1-7.

Anandhakumar, J., and Zeller, W. (2008). Biological control of red stele (*Phytophthora fragariae* var. *fragariae*) and crown rot (*P. cactorum*) disease of strawberry with rhizobacteria. *Journal of Plant Diseases and Protection* **115**(2):49-56.

Angeli, D., Saharan, K., Segarra, G., Sicher, C., and Pertot, I. (2016). Production of *Ampelomyces quisqualis* conidia in submerged fermentation and improvements in the formulation for increased shelf-life. *Crop Protection* **97**(July):135-144.

Archana, S., Manjunath, H., Ranjitham, T. P., Prabakar, K., and Raguchander, T. (2012). Compatibility of azoxystrobin 23 SC with biocontrol agents and insecticides. *Madras Agricultural Journal* **99**(4/6):374-377.

Ash, G. J. (2010). The science, art and business of successful bioherbicides. *Biological Control* **52**(3):230-240.

Bae, Y. S., and Knudsen, G. R. (2005). Soil microbial biomass influence on growth and biocontrol efficacy of *Trichoderma harzianum*. *Biological Control* **32**:236-242.

Balouiri, M., Sadiki, M., and Ibnsouda, S. K. (2016). Methods for *in vitro* evaluating antimicrobial activity: A review. *Journal of Pharmaceutical Analysis* **6**(2):71-79.

Bardin, M., Ajouz, S., Comby, M., Lopez-Ferber, M., Graillot, B., Siegwart, M., and Nicot, P. C. (2015). Is the efficacy of biological control against plant diseases likely to be more durable than that of chemical pesticides? *Frontiers in Plant Science* **6**:566.

Barka, E. A., Gognies, S., Nowak, J., Audran, J.-C., and Belarbi, A. (2002). Inhibitory effect of endophyte bacteria on *Botrytis cinerea* and its influence to promote the grapevine growth. *Biological Control* **24**(2):135-142.

Bashan, Y., de-Bashan, L. E., Prabhu, S. R., and Hernandez, J.-P. (2014). Advances in plant growth-promoting bacterial inoculant technology: formulations and practical perspectives (1998-2013). *Plant and Soil* **378**(1-2):1-33.

Beneduzi, A., Ambrosini, A., and Passaglia, L. M. (2012). Plant growth-promoting rhizobacteria (PGPR): their potential as antagonists and biocontrol agents. *Genetics and Molecular Biology* **35**(4):1044-1051.

Berg, G. (2006). Biological control of fungal soilborne pathogens in strawberries. *In* "Biological Control of Plant Diseases" (S.B. Chincholkar and K. G. Mukerji, eds.), chap. 1, pp. 1-16. Haworth Food & Agricultural Products Press™.

Berg, G., Fritze, A., Roskot, N., and Smalla, K. (2001). Evaluation of potential biocontrol rhizobacteria, from different host plants of *Verticillium dahliae* Kleb. *Journal of Applied Microbiology* **91**(6):963-971.

Berg, G., and Smalla, K. (2009). Plant species and soil type cooperatively shape the structure and function of microbial communities in the rhizosphere. *FEMS Microbiology Ecology* **68**(1):1-13.

Berninger, T., González López, Ó., Bejarano, A., Preininger, C., and Sessitsch, A. (2018). Maintenance and assessment of cell viability in formulation of non-sporulating bacterial inoculants. *Microbial Biotechnology* **11**(2):277-301.

Bhat, R., Geppert, J., Funken, E., and Stamminger, R. (2015). Consumers perceptions and preference for strawberries—A Case study from Germany. *International Journal of Fruit Science* **15**(4):405-424.

Bhattacharjee, R., and Dey, U. (2014). An overview of fungal and bacterial biopesticides to control plant pathogens/diseases. *African Journal of Microbiology Research* **8**(17):1749-1762.

Bilodeau, G. J., Koike, S. T., Uribe, P., and Martin, F. N. (2012). Development of an assay for rapid detection and quantification of *Verticillium dahliae* in soil. *Phytopathology* **102**(3):331-343.

Bisutti, I. L., Steen, S., and Stephan, D. (2013). Does *Metarhizium anisopliae* influence strawberries in presence of pest and disease? *XLVI Annual meetings of the Society for Invertebrate Pathology* in Pittsburgh, Pennsylvania (USA). Conference Proceedings, pp. 11-15.

Bmel (2018). "Ökologischer Landbau in Deutschland. Stand: Januar 2018" Bundesministerium für Ernährung und Landwirtschaft.

Bruck, D. J. (2009). Impact of fungicides on *Metarhizium anisopliae* in the rhizosphere, bulk soil and *in vitro*. *Biocontrol* **54**(4):597-606.

BZFE (2017). Bundeszentrum für Ernährung - Erdbeeren: Verbraucherschutz. www.bzfe.de; Accessed at 17.08.2017

Castro, T., Roggia, S., Wekesa, V. W., de Andrade Moral, R., Demétrio, C. G. B., Delalibera Jr, I., and Klingen, I. (2016). The effect of synthetic pesticides and sulfur used in conventional and organically grown strawberry and soybean on *Neozygites floridana*, a natural enemy of spider mites. *Pest Management Science* **72**(9):1752-1757.

Chandler, D., Bailey, A. S., Tatchell, G. M., Davidson, G., Greaves, J., and Grant, W. P. (2011). The development, regulation and use of biopesticides for integrated pest management. *Philosophical Transactions of the Royal Society B: Biological Sciences* **366**(1573):1987-1998.

Chauhan, H., Bagyaraj, D., Selvakumar, G., and Sundaram, S. (2015). Novel plant growth promoting rhizobacteria—Prospects and potential. *Applied Soil Ecology* **95**:38-53.

Chellemi, D. O., Gamliel, A., Katan, J., and Subbarao, K. V. (2016). Development and deployment of systems-based approaches for the management of soilborne plant pathogens. *Phytopathology* **106**:216-225.

Chojnacka, K. (2015). Innovative bio-products for agriculture. *Open Chemistry* **13**(1):932-937.

Chowdhury, S. P., Dietel, K., Raendler, M., Schmid, M., Junge, H., Borriss, R., Hartmann, A., and Grosch, R. (2013). Effects of *Bacillus amyloliquefaciens* FZB42 on lettuce growth and health under pathogen pressure and its impact on the rhizosphere bacterial community. *PLoS ONE* **8**(7):e68818.

Colla, G., and Rouphael, Y. (2015). Biostimulants in horticulture. *Scientia Horticulturae* **196**:1-2.

Colla, P., Gilardi, G., and Gullino, M. L. (2012). A review and critical analysis of the European situation of soilborne disease management in the vegetable sector. *Phytoparasitica* **40**(5):515-523.

Comby, M., Gacoin, M., Robineau, M., Rabenoelina, F., Ptas, S., Dupont, J., Profizi, C., and Baillieul, F. (2017). Screening of wheat endophytes as biological control agents against Fusarium head blight using two different *in vitro* tests. *Microbiological Research* **202**:11-20.

Commare, R. R., Nandakumar, R., Kandan, A., Suresh, S., Bharathi, M., Raguchander, T., and Samiyappan, R. (2002). *Pseudomonas fluorescens* based bio-formulation for the management of sheath blight disease and leaffolder insect in rice. *Crop Protection* **21**(8):671-677.

Constantinescu, F., Sicuia, O.-A., Fătu, C., Dinu, M. M., Andrei, A.-M., and Mincea, C. (2014). *In vitro* compatibility between chemical and biological products used for seed treatment. *Scientific Papers. Series A. Agronomy* **57**:146-151.

Conti, S., Villari, G., Faugno, S., Melchionna, G., Somma, S., and Caruso, G. (2014). Effects of organic vs. conventional farming system on yield and quality of strawberry grown as an annual or biennial crop in southern Italy. *Scientia Horticulturae* **180**:63-71.

Cray, J. A., Connor, M. C., Stevenson, A., Houghton, J. D. R., Rangel, D. E. N., Cooke, L. R., and Hallsworth, J. E. (2016). Biocontrol agents promote growth of potato pathogens, depending on environmental conditions *Microbial Biotechnology* **9**(3):330-354.

Cray, J. A., Houghton, J. D., Cooke, L. R., and Hallsworth, J. E. (2015). A simple inhibition coefficient for quantifying potency of biocontrol agents against plant-pathogenic fungi. *Biological Control* **81**:93-100.

Daayf, F., Adam, L., and Fernando, W. (2003). Comparative screening of bacteria for biological control of potato late blight (strain US-8), using *in vitro*, detached-leaves, and whole-plant testing systems. *Canadian Journal of Plant Pathology* **25**(3):276-284.

Daryaei, A., Jones, E. E., Ghazalibiglar, H., Glare, T. R., and Falloon, R. E. (2016a). Culturing conditions affect biological control activity of *Trichoderma atroviride* against *Rhizoctonia solani* in ryegrass. *Journal of Applied Microbiology* **121**:461-472.

Daryaei, A., Jones, E. E., Ghazalibiglar, H., Glare, T. R., and Falloon, R. E. (2016b). Effects of temperature, light and incubation period on production, germination and bioactivity of *Trichoderma atroviride*. *Journal of Applied Microbiology* **120**:999-1009.

Daryaei, A., Jones, E. E., Glare, T. R., and Falloon, R. E. (2016c). Biological fitness of *Trichoderma atroviride* during long-term storage, after production in different culture conditions. *Biocontrol Science and Technology* **26**(1):86-103.

Daryaei, A., Jones, E. E., Glare, T. R., and Falloon, R. E. (2016d). pH and water activity in culture media affect biological control activity of *Trichoderma atroviride* against *Rhizoctonia solani*. *Biological Control* **92**:24-30.

Daugaard, H., and Lindhard, H. (2000). Strawberry cultivars for organic production. *Gartenbauwissenschaft* **65**(5):213-217.

de Souza, E. M., Granada, C. E., and Sperotto, R. A. (2016). Plant pathogens affecting the establishment of plant-symbiont interaction. *Frontiers in Plant Science* **7**:15.

de Souza, R., Ambrosini, A., and Passaglia, L. M. (2015). Plant growth-promoting bacteria as inoculants in agricultural soils. *Genetics and Molecular Biology* **38**(4):401-419.

Debode, J., Van Poucke, K., Franca, S. C., Maes, M., Hofte, M., and Heungens, K. (2011). Detection of multiple *Verticillium* species in soil using density flotation and Real-Time Polymerase Chain Reaction. *Plant Disease* **95**(12):1571-1580.

Deketelaere, S., Tyvaert, L., França, S. C., and Höfte, M. (2017). Desirable traits of a good biocontrol agent against Verticillium wilt. *Frontiers in Microbiology* **8**:1186.

Depotter, J., Thomma, B., and Wood, T. (2017). Variable impact of *Verticillium longisporum* on oilseed rape yield in field trials in the United Kingdom. *bioRxiv*:205401.

Destatis (2018). "Land- und Forstwirtschaft, Fischerei: Gemüseerhebung-Anbau und Ernte von Gemüse und Erdbeeren 2017" Statistisches Bundesamt, Fachserie 3 Reihe 3.1.3.

Dickie, G. A., and Bell, C. R. (1995). A full factorial analysis of 9 factors influencing *in vitro* antagonistic screens for potential biocontrol agents. *Canadian Journal of Microbiology* **41**(3):284-293.

Diehl, K., Rebensburg, P., and Lentzsch, P. (2013). Field application of non-pathogenic *Verticillium dahliae* genotypes for regulation of wilt in strawberry plants. *American Journal of Plant Sciences* **4**(7):24-32.

Droby, S., Wisniewski, M., Teixidó, N., Spadaro, D., and Jijakli, M. H. (2016). The science, development, and commercialization of postharvest biocontrol products. *Postharvest Biology and Technology* **122**(December):22-29.

du Jardin, P. (2015). Plant biostimulants: Definition, concept, main categories and regulation. *Scientia Horticulturae* **196**:3-14.

Eikemo, H., Stensvand, A., and Tronsmo, A. (2000). Evaluation of methods of screening strawberry cultivars for resistance to crown rot caused by *Phytophthora cactorum*. *Annals of applied biology* **137**(3):237-244.

Eikemo, H., Stensvand, A., and Tronsmo, A. M. (2003). Induced resistance to *Phytophthora* diseases in strawberry. *Bulletin OILB/SROP* **26**(2):187-192.

Eilenberg, J., Hajek, A., and Lomer, C. (2001). Suggestions for unifying the terminology in biological control. *Biocontrol* **46**(4):387-400.

Elad, Y., Malathrakis, N. E., and Dik, A. J. (1996). Biological control of *Botrytis*-incited diseases and powdery mildews in greenhouse crops. *Crop Protection* **15**(3):229-240.

Elsaid, A., Shehata, S., Abd-El-Moity, T., and Aly, M. (2005). Evaluation of single or combined isolates of *Trichoderma harzianum* in different formulations for controlling root rot diseases of strawberry. *Annals of Agricultural Science (Cairo)* **50**(2):601-611.

Emmert, E. A. B., and Handelsman, J. (1999). Biocontrol of plant disease: a (Gram-) positive perspective. *Fems Microbiology Letters* **171**(1):1-9.

EPPO (2012). Principles of efficacy evaluation for microbial plant protection products. *Bulletin OEPP/EPPO* **42**(3):348-352.

EPPO (2018). European and Mediterranean Plant Protection Organization. *www.eppo.int/QUARANTINE/listA2*.

Esitken, A., Yildiz, H. E., Ercisli, S., Donmez, M. F., Turan, M., and Gunes, A. (2010). Effects of plant growth promoting bacteria (PGPB) on yield, growth and nutrient contents of organically grown strawberry. *Scientia Horticulturae* **124**(1):62-66.

EU (2018). European Commission. https://ec.europa.eu/food/plant/pesticides_en; Accessed 12.02.2018

European Commission (2009a). Directive 2009/128/EC of the European Parliament and of the Council of 21 October 2009 establishing a framework for Community action to achieve the sustainable use of pesticides (1). *Official Journal of the European Union L309;* **52**:71-86.

European Commission (2009b). Regulation (EC) No 1107/2009 of the European Parliament and of the Council of 21 October 2009 concerning the placing of plant protection products on the market and repealing Council Directives 79/117/EEC and 91/414/EEC. *Official Journal of the European Union L309;* **52**:1-50.

European Commission (2014). Review report for the active substance *Metarhizium anisopliae* var. *anisopliae* BIPESCO 5/F52.

Eynck, C., Koopmann, B., Grunewaldt-Stoecker, G., Karlovsky, P., and von Tiedemann, A. (2007). Differential interactions of *Verticillium longisporum* and *V. dahliae* with *Brassica napus* detected with molecular and histological techniques. *European Journal of Plant Pathology* **118**(3):259-274.

FAO (2018). Food and Agriculture Organization of the United Nations http://faostat3.fao.org/; Accessed at 22.03.18

Faraji, S., Shadmehri, A. D., and Mehrvar, A. (2016). Compatibility of entomopathogenic fungi *Beauveria bassiana* and *Metarhizium anisopliae* with some pesticides. *Journal of Entomological Society of Iran* **36**(2):137-146.

Fonseca, F., Marin, M., and Morris, G. (2006). Stabilization of frozen *Lactobacillus delbrueckii* subsp. *bulgaricus* in glycerol suspensions: freezing kinetics and storage temperature effects. *Applied and Environmental Microbiology* **72**(10):6474-6482.

Fonseca, F., Meneghel, J., Cenard, S., Passot, S., and Morris, G. J. (2016). Determination of intracellular vitrification temperatures for unicellular micro-organisms under conditions relevant for cryopreservation. *PLoS ONE* **11**(4):e0152939.

Franks, F. (1994). Accelerated stability testing of bioproducts: attractions and pitfalls. *Trends in Biotechnology* **12**(4):114-117.

Fravel, D. R., Connick Jr, W. J., and Lewis, J. A. (1998). Formulation of microorganisms to control plant diseases. *In* "Formulation of microbial biopesticides" (H. D. Burges, ed.), pp. 187-202. Springer, Dordrecht.

Freeman, S., Minz, D., Kolesnik, I., Barbul, O., Zveibil, A., Maymon, M., Nitzani, Y., Kirshner, B., Rav-David, D., Bilu, A., Dag, A., Shafir, S., and Elad, Y. (2004). *Trichoderma* biocontrol of *Colletotrichum acutatum* and *Botrytis cinerea* and survival in strawberry. *European Journal of Plant Pathology* **110**(4):361-370.

Fuchs, J. G., Moënne-Loccoz, Y., and Défago, G. (2000). The laboratory medium used to grow biocontrol *Pseudomonas* sp Pf153 influences its subsequent ability to protect cucumber from black root rot. *Soil Biology & Biochemistry* **32**(3):421-424.

Gade, R., Chaithanya, B., and Khurade, K. (2014). A comparitive study of different carriers for shelf life of *Pseudomonas fluorescens*. *The Bioscan* **9**(1):287-290.

Gade, R. M., and Armarkar, S. V. (2011). Growth promotion and disease suppression ability of *Pseudomonas fluorescens* in acid lime. *Archives of Phytopathology and Plant Protection* **44**(10):943-950.

Gamliel, A. (2010). Application aspects of integrated pest management. *Journal of Plant Pathology* **92**(S.4):23-26.

Gao, X. N., He, Q. R., Jiang, Y., and Huang, L. L. (2016). Optimization of nutrient and fermentation parameters for antifungal activity by *Streptomyces lavendulae* Xjy and its biocontrol efficacies against *Fulvia fulva* and *Botryosphaeria dothidea*. *Journal of Phytopathology* **164**(3):155-165.

Gardener, B. B. M. (2007). Diversity and ecology of biocontrol *Pseudomonas* spp. in agricultural systems. *Phytopathology* **97**(2):221-226.

Gharbi, Y., Barkallah, M., Bouazizi, E., Cheffi, M., Krid, S., Triki, M. A., and Gdoura, R. (2016). Development and validation of a new real-time assay for the quantification of *Verticillium dahliae* in the soil: a comparison with conventional soil plating. *Mycological Progress* **15**(6):1-13.

Ghasemi, S., and Ahmadzadeh, M. (2013). Optimisation of a cost-effective culture medium for the large-scale production of *Bacillus subtilis* UTB96. *Archives of Phytopathology and Plant Protection* **46**(13):1552-1563.

Gordon, T. R., Shaw, D. V., and Larson, K. A. (2005). Comparative response of strawberries to conidial root-dip inoculations and infection by soilborne microsclerotia of *Verticillium dahliae* Kleb. *Hortscience* **40**(5):1398-1400.

Goud, J. C., and Termorshuizen, A. J. (2003). Quality of methods to quantify microsclerotia of *Verticillium dahliae* in soil. *European Journal of Plant Pathology* **109**(6):523-534.

Goud, J. K. C., Termorshuizen, A. J., Blok, W. J., and van Bruggen, A. H. C. (2004). Long-term effect of biological soil disinfestation on *Verticillium* wilt. *Plant Disease* **88**(7):688-694.

Grabke, A., and Stammler, G. (2015). A *Botrytis cinerea* population from a single strawberry field in Germany has a complex fungicide resistance pattern. *Plant Disease* **99**(8):1078-1086.

Gressel, J. (2015). Are integrated pest management (IPM) and resistance management synonymous o antagonistic? *Pest Management Science* **71**:329-330.

Gullino, M. L. (2005). Environmental impact and risk analysis of bacterial and fungal biocontrol agents - Guest editorial. *Phytoparasitica* **33**(1):3-6.

Guttridge, C. G. (1959). Further evidence for a growth-promoting and flower-inhibiting hormone in strawberry. *Annals of Botany* **23**(4):612-621.

Haas, D., and Défago, G. (2005). Biological control of soil-borne pathogens by fluorescent pseudomonads. *Nature Reviews Microbiology* **3**(4):307-319.

Hancock, J. F. (1999). "Strawberries" CABI Pub, Wallingford, UK.

Harman, G. E. (2006). Overview of mechanisms and uses of *Trichoderma* spp. *Phytopathology* **96**(2):190-194.

Harris, D. C., and Yang, J. R. (1996). The relationship between the amount of *Verticillium dahliae* in soil and the incidence of strawberry wilt as a basis for disease risk prediction. *Plant Pathology* **45**(1):106-114.

Hartmann, A., Schmid, M., van Tuinen, D., and Berg, G. (2009). Plant-driven selection of microbes. *Plant and Soil* **321**(1-2):235-257.

He, L., Xu, Y.-Q., and Zhang, X.-H. (2008). Medium factor optimization and fermentation kinetics for phenazine-1-carboxylic acid production by *Pseudomonas* sp M18G. *Biotechnology and Bioengineering* **100**(2):250-259.

He, Y., Peng, Y., Wu, Z., Han, Y., and Dang, Y. (2015). Survivability of *Pseudomonas putida* RS-198 in liquid formulations and evaluation its growth-promoting abilities on cotton. *Journal of Animal and Plant Sciences* **25**(S.1):180-189.

Heckly, R. J. (1985). Principles of preseving bacteria by freeze-drying. *Developments in Industrial Microbiology* **26**:379-396.

Helbig, J. (2002). Ability of the antagonistic yeast *Cryptococcus albidus* to control *Botrytis cinerea* in strawberry. *BioControl* **47**(1):85-99.

Hernández, A., Weekers, F., Mena, J., Borroto, C., and Thonart, P. (2006). Freeze-drying of the biocontrol agent *Tsukamurella paurometabola* C-924 Predicted stability of formulated powders. *Industrial Biotechnology* **2**(3):209-212.

Hjeljord, L. G., Stensvand, A., and Tronsmo, A. (2001). Antagonism of nutrient-activated conidia of *Trichoderma harzianum* (*atroviride*) P1 against *Botrytis cinerea*. *Phytopathology* **91**(12):1172-1180.

Hjeljord, L. G., and Strømeg, G. M. (2004). Biological control of *Botrytis*: searching for a realistic screen. *Thirteen International Botrytis Symposium* in Antalya (Turkey). Conference Proceedings, pp. 63.

Hol, W. G., Bezemer, T. M., and Biere, A. (2013). Getting the ecology into interactions between plants and the plant growth-promoting bacterium *Pseudomonas fluorescens*. *Induced Plant Responses to Microbes and Insects* **4**(A.81):1-8.

Hynes, R. K., and Boyetchko, S. M. (2006). Research initiatives in the art and science of biopesticide formulations. *Soil Biology & Biochemistry* **38**(4):845-849.

Ikeda, K., Banno, S., Furusawa, A., Shibata, S., Nakaho, K., and Fujimura, M. (2015). Crop rotation with broccoli suppresses *Verticillium* wilt of eggplant. *Journal of General Plant Pathology* **81**(1):77-82.

Jackson, M. A. (1997). Optimizing nutritional conditions for the liquid culture production of effective fungal biological control agents. *Journal of Industrial Microbiology & Biotechnology* **19**(3):180-187.

Jain, A., and Das, S. (2016). Insight into the interaction between plants and associated fluorescent *Pseudomonas* spp. *International Journal of Agronomy* **2016**:4269010.

Jensen, D. F., Karlsson, M., Sarrocco, S., and Vannacci, G. (2016). Biological control using microorganisms as an alternative to disease resistance. *In* "Plant Pathogen Resistance Biotechnology" (D. E. Collinge, ed.), chap. 18, pp. 341-363. John Wiley & Sons.

Ji, P., Campbell, H. L., Kloepper, J. W., Jones, J. B., Suslow, T. V., and Wilson, M. (2006). Integrated biological control of bacterial speck and spot of tomato under field conditions using foliar biological control agents and plant growth-promoting rhizobacteria. *Biological Control* **36**(3):358-367.

Jones, K. A., and Burges, H. D. (1998). Technology of formulation and application. *In* "Formulation of Microbial Biopesticides" (H. D. Burges, ed.), pp. 7-30. Springer, Dordrecht.

Kefi, A., Slimene, I. B., Karkouch, I., Rihouey, C., Azaeiz, S., Bejaoui, M., Belaid, R., Cosette, P., Jouenne, T., and Limam, F. (2015). Characterization of endophytic *Bacillus strains* from tomato plants (*Lycopersicon esculentum*) displaying antifungal activity against *Botrytis cinerea* Pers. *World Journal of Microbiology and Biotechnology* **31**(12):1967-1976.

Keswani, C., Bisen, K., Singh, V., Sarma, B. K., and Singh, H. B. (2016). Formulation technology of biocontrol agents: Present status and future prospects. *In* "Bioformulations: for Sustainable Agriculture" (M. S. Arora N., Balestrini R., ed.), pp. 35-52. Springer, New Delhi.

Khan, M. R., Anwer, M. A., and Shahid, S. (2011). Management of gray mold of chickpea, *Botrytis cinerea* with bacterial and fungal biopesticides using different modes of inoculation and application. *Biological Control* **57**(1):13-23.

Khan, M. R., Brien, E. O., Carney, B. F., and Doohan, F. M. (2010). A fluorescent pseudomonad shows potential for the control of net blotch disease of barley. *Biological Control* **54**(1):41-45.

King, V. A.-E., Lin, H.-J., and Liu, C.-F. (1998). Accelerated storage testing of freeze-dried and controlled low-temperature vacuum dehydrated *Lactobacillus acidophilus*. *The Journal of General and Applied Microbiology* **44**(2):161-165.

Kishore, K. G., Pande, S., and Podile, A. R. (2005). Management of late leaf spot of groundnut (*Arachis hypogaea*) with chlorothalonil-tolerant isolates of *Pseudomonas aeruginosa*. *Plant Pathology* **54**(3):401-408.

Klingen, I., and Westrum, K. (2007). The effect of pesticides used in strawberries on the phytophagous mite *Tetranychus urticae* (Acari: Tetranychidae) and its fungal natural enemy *Neozygites floridana* (Zygomycetes: Entomophthorales). *Biological Control* **43**(2):222-230.

Klosterman, S. J., Atallah, Z. K., Vallad, G. E., and Subbarao, K. V. (2009). Diversity, pathogenicity, and management of *Verticillium* species. *Annual Review of Phytopathology* **47**:39-62.

Knudsen, G. R., and Dandurand, L.-M. C. (2014). Ecological complexity and the success of fungal biological control agents. *Advances in Agriculture* **2014**:502703.

Knudsen, I., Hockenhull, J., Jensen, D. F., Gerhardson, B., Hökeberg, M., Tahvonen, R., Teperi, E., Sundheim, L., and Henriksen, B. (1997). Selection of biological control agents for controlling soil and seed-borne diseases in the field. *European Journal of Plant Pathology* **103**(9):775-784.

Köhl, J., Postma, J., Nicot, P. C., Ruocco, M., and Blum, B. (2011). Stepwise screening of microorganisms for commercial use in biological control of plant-pathogenic fungi and bacteria. *Biological Control* **57**:1-12.

Koike, S. T., and Gordon, T. G. (2015). Management of Fusarium wilt of strawberry. *Crop Protection* **73**:67-72.

Kuchta, P., Jecz, T., and Korbin, M. (2008). The suitability of PCR-based techniques for detecting *Verticillium dahliae* in strawberry plants and soil. *Journal of Fruit and Ornamental Plant Research* **16**:295-304.

Kumar, P., and Mane, S. (2017). Studies on the compatibility of biocontrol agents with certain fungicides. *International Journal of Current Microbiology and Applied Sciences* **6**(3):1639-1644.

Kurze, S., Bahl, H., Dahl, R., and Berg, G. (2001). Biological control of fungal strawberry diseases by *Serratia plymuthica* HRO-C48. *Plant Disease* **85**(5):529-534.

Lahlali, R., Peng, G., Gossen, B., McGregor, L., Yu, F., Hynes, R., Hwang, S., McDonald, M., and Boyetchko, S. (2013). Evidence that the biofungicide Serenade (*Bacillus subtilis*) suppresses clubroot on canola via antibiosis and induced host resistance. *Phytopathology* **103**(3):245-254.

Lamichhane, J. R., Bischoff-Schaefer, M., Bluemel, S., Dachbrodt-Saaydeh, S., Dreux, L., Jansen, J. P., Kiss, J., Köhl, J., Kudsk, P., and Malausa, T. (2017). Identifying obstacles and ranking common biological control research priorities for Europe to manage most economically important pests in arable, vegetable and perennial crops. *Pest Management Science* **73**(1):14-21.

Lefebvre, M., Langrell, S. R., and Gomez-y-Paloma, S. (2015). Incentives and policies for integrated pest management in Europe: a review. *Agronomy for Sustainable Development* **35**(1):27-45.

Liu, Z., Wei, H., Li, Y., Li, S., Luo, Y., Zhang, D., and Ni, L. (2014). Optimization of the spray drying of a *Paenibacillus polymyxa*-based biopesticide on pilot plant and production scales. *Biocontrol Science and Technology* **24**(4):426-435.

Maas, J. L. (1998). "Compendium of Strawberry Diseases" APS Press, St. Paul, MN.

Madhavi, G. B., Bhattiprolu, S. L., and Reddy, V. B. (2011). Compatibility of biocontrol agent *Trichoderma viride* with various pesticides. *Journal of Horticultural Sciences* 6(1):71-73.

Manikandan, R., Saravanakumar, D., Rajendran, L., Raguchander, T., and Samiyappan, R. (2010). Standardization of liquid formulation of *Pseudomonas fluorescens* Pf1 for its efficacy against *Fusarium* wilt of tomato. *Biological Control* 54(2):83-89.

Martin, F. N. (2003). Development of alternative strategies for management of soilborne pathogens currently controlled with methyl bromide. *Annual Review of Phytopathology* 41:325-350.

Martin, F. N., and Bull, C. T. (2002). Biological approaches for control of root pathogens of strawberry. *Phytopathology* 92(12):1356-1362.

Massart, S., Martinez-Medina, M., and Jijakli, M. H. (2015a). Biological control in the microbiome era: Challenges and opportunities. *Biological Control* 89:98-108.

Massart, S., Perazzolli, M., Höfte, M., Pertot, I., and Jijakli, M. H. (2015b). Impact of the omic technologies for understanding the modes of action of biological control agents against plant pathogens. *BioControl* 60(6):725-746.

Mathivanan, N., Prabavathy, V. R., and Vijayanandraj, V. R. (2005). Application of talc formulations of *Pseudomonas fluorescens* migula and *Trichoderma viride* pers. Ex SF Gray decrease the sheath blight disease and enhance the plant growth and yield in rice. *Journal of Phytopathology* 153(11-12):697-701.

Matyjaszczyk, E. (2015). Products containing microorganisms as a tool in integrated pest management and the rules of their market placement in the European Union. *Pest Management Science* 71:1201-1206.

Maurya, M. K., Singh, R., and Tomer, A. (2014). *In vitro* evaluation of antagonistic activity of *Pseudomonas fluorescens* against fungal pathogen. *Journal of Biopesticides* 7(1):43-46.

Miyamoto-Shinohara, Y., Imaizumi, T., Sukenobe, J., Murakami, Y., Kawamura, S., and Komatsu, Y. (2000). Survival rate of microbes after freeze-drying and long-term storage. *Cryobiology* 41(3):251-255.

Mizrahi, S. (2004). Accelerated shelf-life tests. *In* "Understanding and measuring the shelf-life of food" (D. K. a. P. Subramaniam, ed.), chap. 5, pp. 317-337. Woodhead Publishing Limited.

Mnif, I., and Ghribi, D. (2015). Potential of bacterial derived biopesticides in pest management. *Crop Protection* 72:52-64.

Mohiddin, F. A., and Khan, M. R. (2013). Tolerance of fungal and bacterial biocontrol agents to six pesticides commonly used in the control of soil borne plant pathogens. *African Journal of Agricultural Research* 8(43):5331-5334.

Mol, L., Huisman, O. C., Scholte, K., and Struik, P. C. (1996). Theoretical approach to the dynamics of the inoculum density of *Verticillium dahliae* in the soil: first test of a simple model. *Plant Pathology* 45(2):192-204.

Montesinos, E. (2003). Development, registration and commercialization of microbial pesticides for plant protection. *International Microbiology* 6(4):245-252.

Morgan, C., and Vesey, G. (2009). Freeze-drying of microorganisms. *In* "Encyclopedia of microbiology" (M. Schaechter, ed.), pp. 162-173. Elsevier B.V.

Morgan, C. A., Herman, N., White, P. A., and Vesey, G. (2006). Preservation of micro-organisms by drying: A review. *Journal of Microbiological Methods* 66(2):183-193.

Moročko-Bičevska, I., and Fatehi, J. (2011). Infection and colonization of strawberry by *Gnomonia fragariae* strain expressing green fluorescent protein. *European Journal of Plant Pathology* 129(4):567-577.

Moser, R., Pertot, I., Elad, Y., and Raffaelli, R. (2008). Farmers' attitudes toward the use of biocontrol agents in IPM strawberry production in three countries. *Biological Control* **47**(2):125-132.

Mota, M. S., Gomes, C. B., Júnior, I. T. S., and Moura, A. B. (2017). Bacterial selection for biological control of plant disease: criterion determination and validation. *Brazilian Journal of Microbiology* **48**(1):62-70.

Mputu, K., and Thonart, P. (2013). Optimisation of production, freeze-drying and storage of *Pseudomonas fluorescens* BTP1. *International Journal of Microbiology Research* **5**(2):370-373.

Mputu, K. J.-N. (2014). Impact du séchage sur la viabilité de *Pseudomonas fluorescens* (synthèse bibliographique). *Biotechnologie, Agronomie, Société et Environnement= Biotechnology, Agronomy, Society and Environment* **18**(1):134-141.

Mputu, K. J.-N., Destain, J., Noki, P., and Thonart, P. (2012a). Accelerated storage testing of freeze-dried *Pseudomonas fluorescens* BTP1, BB2 and PI9 strains. *African Journal of Biotechnology* **11**(95):16187-16191.

Mputu, K. J.-N., Pierart, C., Weekers, F., Destain, J., and Thonart, P. (2012b). Impact of protective compounds on the viability, physiological state and lipid degradation of freeze-dried *Pseudomonas fluorescens* BTP1 during storage. *International Journal of Biotechnology and Biochemistry* **8**(4):17-26.

Mukherjee, A., and Babu, S. P. S. (2013). *Pseudomonas fluorescens* mediated suppression of *Meloidogyne incognita* infection of cowpea and tomato. *Archives of Phytopathology and Plant Protection* **46**(5):607-616.

Muller, A., Schader, C., Scialabba, N. E.-H., Brüggemann, J., Isensee, A., Erb, K.-H., Smith, P., Klocke, P., Leiber, F., and Stolze, M. (2017). Strategies for feeding the world more sustainably with organic agriculture. *Nature Communications* **8**:1290.

Müller, H., and Berg, G. (2008). Impact of formulation procedures on the effect of the biocontrol agent *Serratia plymuthica* HRO-C48 on Verticillium wilt in oilseed rape. *Biocontrol* **53**(6):905-916.

Muñoz-Rojas, J., Bernal, P., Duque, E., Godoy, P., Segura, A., and Ramos, J. L. (2006). Involvement of cyclopropane fatty acids in the response of *Pseudomonas putida* KT2440 to freeze-drying. *Applied and Environmental Microbiology* **72**(1):472-477.

Myresiotis, C. K., Vryzas, Z., and Papadopoulou-Mourkidou, E. (2012). Biodegradation of soil-applied pesticides by selected strains of plant growth-promoting rhizobacteria (PGPR) and their effects on bacterial growth. *Biodegradation* **23**(2):297-310.

Nagarajkumar, M., Bhaskaran, R., and Velazhahan, R. (2004). Involvement of secondary metabolites and extracellular lytic enzymes produced by *Pseudomonas fluorescens* in inhibition of *Rhizoctonia solani*, the rice sheath blight pathogen. *Microbiological Research* **159**(1):73-81.

Nandakumar, R., Babu, S., Viswanathan, R., Raguchander, T., and Samiyappan, R. (2001). Induction of systemic resistance in rice against sheath blight disease by *Pseudomonas fluorescens*. *Soil Biology & Biochemistry* **33**(4-5):603-612.

Nehra, V., and Choudhary, M. (2015). A review on plant growth promoting rhizobacteria acting as bioinoculants and their biological approach towards the production of sustainable agriculture. *Journal of Applied and Natural Science* **7**(1):540-556.

Neubauer, C., and Heitmann, B. (2011). Quantitative detection of *Verticillium dahliae* in soil as a basis for selection of planting sites in horticulture. *Journal fur Kulturpflanzen* **63**(1):1-8.

OECD (2018). The Organisation for Economic Co-operation and Development. http://www.oecd.org/chemicalsafety/pesticides-biocides/biological-pesticides.htm; Accessed 12.02.2018

Olbricht, K., and Hanke, M.-V. (2008). Strawberry breeding for disease resistance in Dresden. *Ecofruit-13th International Conference on Cultivation Technique and Phytopathological Problems in Organic Fruit-Growing* in Weinsberg (Germany). Conference Proceedings, pp. 144-147.

Paau, A. S. (1988). Formulations useful in applying beneficial microorganisms to seed. *Trends in Biotechnology* **6**(11):276-279.

Palazzini, J., Ramirez, M., Alberione, E., Torres, A., and Chulze, S. (2009). Osmotic stress adaptation, compatible solutes accumulation and biocontrol efficacy of two potential biocontrol agents on Fusarium head blight in wheat. *Biological Control* **51**(3):370-376.

Palazzini, J. M., Alberione, E., Torres, A., Donat, C., Köhl, J., and Chulze, S. (2016). Biological control of *Fusarium graminearum* sensu stricto, causal agent of Fusarium head blight of wheat, using formulated antagonists under field conditions in Argentina. *Biological Control* **94**:56-61.

Palmfeldt, J., Rådström, P., and Hahn-Hägerdal, B. (2003). Optimisation of initial cell concentration enhances freeze-drying tolerance of *Pseudomonas chlororaphis*. *Cryobiology* **47**(1):21-29.

Parnell, J. J., Berka, R., Young, H. A., Sturino, J. M., Kang, Y. W., Barnhart, D. M., and DiLeo, M. V. (2016). From the Lab to the Farm: An Industrial Perspective of Plant Beneficial Microorganisms. *Frontiers in Plant Science* **7**:1110.

Pascual, I., Azcona, I., Morales, F., Aguirreolea, J., and Sanchez-Diaz, M. (2010). Photosynthetic response of pepper plants to wilt induced by *Verticillium dahliae* and soil water deficit. *Journal of Plant Physiology* **167**(9):701-708.

Pastrana, A. M., Basallote-Ureba, M. J., Aguado, A., Akdi, K., and Capote, N. (2016). Biological control of strawberry soil-borne pathogens *Macrophomina phaseolina* and *Fusarium solani*, using *Trichoderma asperellum* and *Bacillus* spp. *Phytopathologia Mediterranea* **55**(1):109-120.

Pertot, I., Zasso, R., Amsalem, L., Baldessari, M., Angeli, G., and Elad, Y. (2008). Integrating biocontrol agents in strawberry powdery mildew control strategies in high tunnel growing systems. *Crop Protection* **27**(3):622-631.

Pliego, C., Ramos, C., de Vicente, A., and Cazorla, F. M. (2011). Screening for candidate bacterial biocontrol agents against soilborne fungal plant pathogens. *Plant and Soil* **340**(1-2):505-520.

Porras, M., Barran, C., Arroyo, F. T., Santos, B., Blanco, C., and Romero, F. (2007). Reduction of *Phytophthora cactorum* in strawberry fields by *Trichoderma* spp. and soil solarization. *Plant Disease* **91**(2):142-146.

Prasad, J., McJarrow, P., and Gopal, P. (2003). Heat and osmotic stress responses of probiotic *Lactobacillus rhamnosus* HN001 (DR20) in relation to viability after drying. *Applied and Environmental Microbiology* **69**(2):917-925.

Rabindran, R., and Vidhyasekaran, P. (1996). Development of a formulation of *Pseudomonas fluorescens* PfALR2 for management of rice sheath blight. *Crop Protection* **15**(8):715-721.

Reganold, J. P., Andrews, P. K., Reeve, J. R., Carpenter-Boggs, L., Schadt, C. W., Alldredge, J. R., Ross, C. F., Davies, N. M., and Zhou, J. (2010). Fruit and soil quality of organic and conventional strawberry agroecosystems. *PLoS ONE* **5**(9):e12346.

Reganold, J. P., and Wachter, J. M. (2016). Organic agriculture in the twenty-first century. *Nature Plants* **2**(2):1-8.

Robinson-Boyer, L., Jeger, M. J., Xu, X.-M., and Jeffries, P. (2009). Management of strawberry grey mould using mixtures of biocontrol agents with different mechanisms of action. *Biocontrol Science and Technology* **19**(10):1051-1065.

Ruzzi, M., and Aroca, R. (2015). Plant growth-promoting rhizobacteria act as biostimulants in horticulture. *Scientia Horticulturae* **196**:124-134.

Saxena, D., Tewari, A. K., and Dinesh, R. (2014). The in vitro effect of some commonly used fungicides, insecticides and herbicides for their compatibility with *Trichoderma harzianum* PBT23. *World Applied Sciences Journal* **31**(4):444-448.

Schisler, D., Slininger, P., and Olsen, N. (2016). Appraisal of selected osmoprotectants and carriers for formulating Gram-negative biocontrol agents active against Fusarium dry rot on potatoes in storage. *Biological Control* **98**:1-10.

Schisler, D. A., Slininger, R. J., Behle, R. W., and Jackson, M. A. (2004). Formulation of *Bacillus* spp. for biological control of plant diseases. *Phytopathology* **94**(11):1267-1271.

Seufert, V., Ramankutty, N., and Foley, J. A. (2012). Comparing the yields of organic and conventional agriculture. *Nature* **485**(7397):229-232.

Shehata, H. R., Lyons, E. M., Jordan, K. S., and Raizada, M. N. (2016). Relevance of *in vitro* agar based screens to characterize the anti-fungal activities of bacterial endophyte communities. *BMC Microbiology* **16**(8).

Shen, T., Wang, C., Yang, H., Deng, Z., Wang, S., Shen, B., and Shen, Q. (2016). Identification, solid-state fermentation and biocontrol effects of *Streptomyces hygroscopicus* B04 on strawberry root rot. *Applied Soil Ecology* **103**:36-43.

Shennan, C., Muramoto, J., Lamers, J., Mazzola, M., Rosskopf, E., Kokalis-Burelle, N., Momma, N., Butler, D., and Kobara, Y. (2014). Anaerobic soil disinfestation for soil borne disease control in strawberry and vegetable systems: current knowledge and future directions. *Acta Horticulturae* **1044**:165-175.

Shetty, K. G., Subbarao, K. V., Huisman, O. C., and Hubbard, J. C. (2000). Mechanism of broccoli-mediated Verticillium wilt reduction in cauliflower. *Phytopathology* **90**(3):305-310.

Short, D. P., Sandoya, G., Vallad, G. E., Koike, S. T., Xiao, C.-L., Wu, B.-M., Gurung, S., Hayes, R. J., and Subbarao, K. V. (2015). Dynamics of *Verticillium* species microsclerotia in field soils in response to fumigation, cropping patterns, and flooding. *Phytopathology* **105**(5):638-645.

Shu, G., Zhang, B., Hui, Y., Chen, H., and Wan, H. (2017). Optimization of cryoprotectants for *Streptococcus thermophilus* during freeze-drying using Box-Behnken experimental design of response surface methodology. *Emirates Journal of Food and Agriculture* **29**(4):256-263.

Siegwart, M., Graillot, B., Lopez, C. B., Besse, S., Bardin, M., Nicot, P. C., and Lopez-Ferber, M. (2015). Resistance to bio-insecticides or how to enhance their sustainability: a review. *Frontiers in Plant Science* **6**:381.

Singh, A., and Dubey, S. (2017). Transgenic plants and soil microbes. *In* "Current Developments in Biotechnology and Bioengineering: Crop Modification, Nutrition, and Food Production" (S. D. A. P. R. Sangwan, ed.), chap. 8, pp. 163-185. Elsevier B.V.

Singh, B., and Dubey, S. C. (2010). Bioagent based integrated management of Phytophthora blight of pigeonpea. *Archives of Phytopathology and Plant Protection* **43**(9):922-929.

Singh, H. B., and Singh, D. P. (2009). From biological control to bioactive metabolites: prospects with *Trichoderma* for safe human food. *Pertanika Journal of Tropical Agricultural Science* **32**(1):99-110.

Singh, R., and Arora, N. K. (2016). Bacterial formulations and delivery systems against pests in sustainable agro-food production. *In* "Reference Module in Food Sciences", pp. 1-11. Academic Press, Elsevier.

Slininger, P. J., and Shea-Wilbur, M. A. (1995). Liquid-culture pH, temperature, and carbon (not nitrogen) source regulate phenazine productivity of take-all biocontrol agent *Pseudomonas fluorescens* 2-79. *Applied Microbiology and Biotechnology* **43**(5):794-800.

Smith, D., and Onions, A. H. S. (1983). "The preservation and maintenance of living fungi" Commonwealth Mycological Institute, Kew, UK.

Spadaro, D., and Gullino, M. L. (2005). Improving the efficacy of biocontrol agents against soilborne pathogens. *Crop Protection* **24**(7):601-613.

Steffek, R., Spornberger, A., and Altenburger, J. (2006). Detection of microsclerotia of *Verticillium dahliae* in soil samples and prospects to reduce the inoculum potential of the fungus in the soil. *Agriculturae Conspectus Scientificus* **71**(4):145-148.

Stephan, D., Da Silva, A. P. M., and Bisutti, I. L. (2016). Optimization of a freeze-drying process for the biocontrol agent *Pseudomonas* spp. and its influence on viability, storability and efficacy. *Biological Control* **94**:74-81.

Stephan, D., Matos da Silva, A. P., and Bisutti, I. L. (2007). Development of a freeze-drying technique for *Pseudomonas fluorescens* strain Pf153. *Jahrestagung für Allgemeinen und Angewandte Mikrobiologie (VAAM)* in Jena (Germany). Biospektrum Sonderausgabe, pp. 197.

Sultan, N. S., Raipat, B. S., and Sinha, M. (2013). Isolation and characterization of *Bacillus cereus* strain JY9 and *Methylobacterium* sp. HJM27 and their growth kinetics studies in presence of pesticides. *Journal of Biopesticides* **6**(1):26-31.

Tahmatsidou, V., O'Sullivan, J., Cassells, A. C., Voyiatzis, D., and Paroussi, G. (2006). Comparison of AMF and PGPR inoculants for the suppression of *Verticillium* wilt of strawberry (*Fragaria* x *ananassa* cv. Selva). *Applied Soil Ecology* **32**(3):316-324.

Tanimomo, J., Delcenserie, V., Taminiau, B., Daube, G., Saint-Hubert, C., and Durieux, A. (2016). Growth and freeze-drying optimization of *Bifidobacterium crudilactis*. *Food and Nutrition Sciences* **7**(7):616-626.

Teixidó, N., Canamas, T. P., Abadias, M., Usall, J., Solsona, C., Casals, C., and Vinas, I. (2006). Improving low water activity and desiccation tolerance of the biocontrol agent *Pantoea agglomerans* CPA-2 by osmotic treatments. *Journal of Applied Microbiology* **101**(4):927-937.

Tiwari, and Tripathi (2014). The multifaceted role of the *Trichoderma* system in biocontrol. *In* "Biological Controls for Preventing Food Deterioration: Strategies for Pre- and Postharvest Management" (N. Sharma, ed.), chap. 9, pp. 183-210. John Wiley & Sons.

Tjamos, E. C., Tjamos, S. E., and Antoniou, P. P. (2010). Biological management of plant diseases: Highlights on research and application. *Journal of Plant Pathology* **92**(4):17-21.

Trabelsi, D., and Mhamdi, R. (2013). Microbial inoculants and their impact on soil microbial communities: A review. *Biomed Research International* **11**:863240.

Uppal, A. K., El Hadrami, A., Adam, L. R., Tenuta, M., and Daayf, F. (2008). Biological control of potato Verticillium wilt under controlled and field conditions using selected bacterial antagonists and plant extracts. *Biological Control* **44**(1):90-100.

Vacheron, J., Moënne-Loccoz, Y., Dubost, A., Gonçalves-Martins, M., Muller, D., and Prigent-Combaret, C. (2016). Fluorescent *Pseudomonas* strains with only few plant-beneficial properties are favored in the maize rhizosphere. *Frontiers in Plant Science* **7**:1212.

Vassilev, N., Vassileva, M., Lopez, A., Martos, V., Reyes, A., Maksimovic, I., Eichler-Lobermann, B., and Malusa, E. (2015). Unexploited potential of some biotechnological techniques for biofertilizer production and formulation. *Applied Microbiology and Biotechnology* **99**(12):4983-4996.

Venturi, V., and Keel, C. (2016). Signaling in the rhizosphere. *Trends in Plant Science* **21**(3):187-198.

Verbon, E. H., and Liberman, L. M. (2016). Beneficial microbes affect endogenous mechanisms controlling root development. *Trends in Plant Science* **21**(3):218-229.

Vestberg, M., Kukkonen, S., Saari, K., Parikka, P., Huttunen, J., Tainio, L., Devos, N., Weekers, F., Kevers, C., Thonart, P., Lemoine, M. C., Cordier, C., Alabouvette, C., and Gianinazzi, S. (2004). Microbial inoculation for improving the growth and health of micropropagated strawberry. *Applied Soil Ecology* **27**(3):243-258.

Vidhyasekaran, P., Rabindran, R., Muthamilan, M., Nayar, K., Rajappan, K., Subramanian, N., and Vasumathi, K. (1997). Development of a powder formulation of *Pseudomonas fluorescens* for control of rice blast. *Plant Pathology* **46**(3):291-297.

Vimi, L., Sindu, P. G., Jiphy Jose, P., and Pushpalatha, P. B. (2016). Compatibility of *Pseudomonas fluorescens* with fungicides used in banana cultivation. *International Journal of Agriculture Innovations and Research* **5**(3):487-488.

Wei, F., Hu, X., and Xu, X. (2016). Dispersal of *Bacillus subtilis* and its effect on strawberry phyllosphere microbiota under open field and protection conditions. *Scientific Reports* **6**:22611.

Wei, Z., Huang, J., Yang, C., Xu, Y., Shen, Q., and Chen, W. (2015). Screening of suitable carriers for *Bacillus amyloliquefaciens* strain QL-18 to enhance the biocontrol of tomato bacterial wilt. *Crop Protection* **75**:96-103.

Wessendorf, J., and Lingens, F. (1989). Effect of culture and soil conditions on survival of *Pseudomonas fluorescens* R1 in soil. *Applied Microbiology and Biotechnology* **31**(1):97-102.

Yang, W., Zheng, L., Liu, H. X., Wang, K. B., Yu, Y. Y., Luo, Y. M., and Guo, J. H. (2014). Evaluation of the effectiveness of a consortium of three plant-growth promoting rhizobacteria for biocontrol of cotton Verticillium wilt. *Biocontrol Science and Technology* **24**(5):489-502.

Yim, B., Hanschen, F. S., Wrede, A., Nitt, H., Schreiner, M., Smalla, K., and Winkelmann, T. (2016). Effects of biofumigation using *Brassica juncea* and *Raphanus sativus* in comparison to disinfection using Basamid on apple plant growth and soil microbial communities at three field sites with replant disease. *Plant and Soil* **406**(1):1-20.

Zeise, K., and von Tiedemann, A. (2002). Application of RAPD-PCR for virulence type analysis within *Verticillium dahliae* and *V. longisporum*. *Journal of Phytopathology-Phytopathologische Zeitschrift* **150**(10):557-563.

Zhao, Y., Huo, X., Zhang, Y., and Miao, Z. (2016). Optimization of *Lactobacillus rhamnosus* viability during freeze-drying by Box-Behnken design. *The Annals of the University of Dunarea de Jos of Galati. Fascicle VI. Food Technology* **40**(1):23-31.

Optimization of a freeze-drying process for the biocontrol agent *Pseudomonas* spp. and its influence on viability, storability and efficacy (2016).

Stephan D., Matos da Silva A.-P. and Bisutti I.L.

Biological Control 94: 74-81

http://dx.doi.org/10.1016/j.biocontrol.2015.12.004

Biological Control 94 (2016) 74–81

Contents lists available at ScienceDirect

Biological Control

journal homepage: www.elsevier.com/locate/ybcon

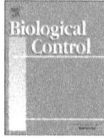

Optimization of a freeze-drying process for the biocontrol agent *Pseudomonas* spp. and its influence on viability, storability and efficacy

Dietrich Stephan *, Ana-Paula Matos Da Silva, Isabella L. Bisutti

Julius Kühn-Institut, Federal Research Centre on Cultivated Plants, Institute for Biological Control, Heinrichstrasse 243, D-64287 Darmstadt, Germany

HIGHLIGHTS

- Freeze-drying is a suitable process for formulation of various *Pseudomonas* strains.
- After optimization the viability after freeze-drying of strain Pf153 was not anymore reduced.
- Particularly CPAs influence the viability, storability and efficacy of pseudomonads.

GRAPHICAL ABSTRACT

ARTICLE INFO

Article history:
Received 30 March 2015
Revised 9 December 2015
Accepted 9 December 2015
Available online 10 December 2015

Keywords:
Alternaria
Biological control
Botrytis cinerea
Lyophilization
Cryoprotective agent

ABSTRACT

Our objective was to investigate the practicability of freeze-drying to formulate and stabilize pseudomonads. Using a not optimized freeze-drying protocol the viability of five *Pseudomonas* strains after freeze-drying ranged only between 2% and 10%. To improve the viability, three different freezing rates, three drying temperatures and 20 cryo-protective agents (CPAs) were compared. The viability after freeze-drying was particularly influenced by the CPAs. Especially, saccharose, glucose, lactose, skimmed milk and ligninosulfonic acid protected vegetative cells within the freeze-drying process. After optimizing the process the viability of the tested *P. fluorescens* strain Pf153 before and after freeze-drying was not anymore reduced. Within storability tests of four *Pseudomonas* strains with the best five CPAs, lactose, saccharose and skimmed milk performed the highest viability.

When the efficacy of freshly produced and freeze-dried pseudomonads was compared in two different plant-pathogen systems, no significant differences were obtained. But CPAs were influencing the efficacy of strain Pf153 against *Botrytis cinerea*. The same efficacy was obtained when strain Pf153 was freeze-dried with lactose or skimmed milk whereas formulation with saccharose had a negative effect.

Our results demonstrate that freeze-drying is a realistic drying technology for pseudomonads. For the development of the best freeze-drying protocol beside viability, the efficacy and storability are important additional selection criteria.

© 2015 Elsevier Inc. All rights reserved.

1. Introduction

Pseudomonads are well described Gram-negative microorganisms that are known as colonizers of the rhizosphere (Walsh et al., 2000; Weller, 2007) and the phyllosphere (Hirano

* Corresponding author.
  E-mail address: dietrich.stephan@jki.bund.de (D. Stephan).

http://dx.doi.org/10.1016/j.biocontrol.2015.12.004
1049-9644/© 2015 Elsevier Inc. All rights reserved.

and Upper, 2000; Krimm et al., 2005). It has been demonstrated that pseudomonads can be used to control plant pathogens (Fang et al., 2007; Johansson and Wright, 2003; O'Callaghan et al., 2006; Someya et al., 2007; Walsh et al., 2000). However, there has been limited success in the development of commercial products containing viable cells of pseudomonads. E.g., in Europe only two *Pseudomonas* strains are approved for use in plant protection (Commission, 2015).

Beside the selection of antagonistic bacterial strains the development of production and formulation technologies of the active microbial ingredients are important steps. Still, the major bottleneck for commercialization of pseudomonads is the availability of appropriate formulations. Although various types of *Pseudomonas* formulations were developed by several authors (Bora et al., 2004; Commare et al., 2002; Dandurand et al., 1994; Mathivanan et al., 2005; Moenne-Loccoz et al., 1999; Rajappan et al., 2002; Senthil et al., 2003), still limited viability of formulated cells and their storability are the most critical points.

Oetjen and Haseley (2004) summarized that one of the oldest methods of preservation is the drying of food and herbs. Freeze-drying or lyophilization was first carried out 1890 and the industrial freeze-drying began with the production of preserved blood plasma and penicillium. Today, freeze-drying is an established method to preserve micro-organisms. The process starts with the freezing step followed by sublimation and desorption of the water in the product. For industrial drying of micro-organisms it is important to obtain maximum survival. The major causes for loss of viability during freeze-drying are osmotic shock and membrane injury (Heckly, 1985). Different micro-organisms and even strains of a given species vary in their viability after freeze-drying (Carvalho et al., 2004b; Heckly, 1985). Additionally, factors such as freezing rate, drying temperature or the cryo-protective agents (CPAs) can influence the viability after freeze-drying (Abadias et al., 2001a; Carvalho et al., 2004a; Clement, 1961). Although the method of freeze-drying is well established for several micro-organisms, specific freeze-drying protocols for commercial development of *Pseudomonas* spp. preparations have not been published.

Therefore, research was carried out to elaborate and optimize freeze-drying protocols for different *Pseudomonas* strains. The effect of different freezing rates and drying temperatures was evaluated, a range of CPAs were compared and their influence on viability, storability and efficacy of the freeze-dried preparations was investigated.

2. Materials and methods

2.1. Bacterial strains

The *P. putida* strain MF416 (NCIMB accession number 40942) was provided by the Swedish University of Agricultural Science in Uppsala and strain I112 was provided by Eckhard Koch of the Julius Kühn-Institut in Darmstadt, Germany. The *P. fluorescens* strain Pf153 and *P. protegens* strain CHA0 were provided by Monika Maurhofer of the Swiss Federal Institute of Technology in Zürich and the *P. chlororaphis* strain PCL1391 by Ben Lugtenberg of the Leiden University, Netherlands.

2.2. Cultivation of bacteria

All pseudomonads were stored at −80 °C in Cryo-tubes (Microbank, Pro-Lab Diagnostics, Ontario, Canada) and routinely cultured at 25 °C on Tryptic Soy Agar (TSA) containing 0.3% (w/v) Tryptic Soy Broth (TSB, Difco, Germany) and 1.5% agar-agar. For preparation of pre-cultures, 25 ml of Nutrient broth (NB, CM 1,

Oxoid, Germany) in 50-ml Erlenmeyer flasks were autoclaved at 121 °C for 20 min, inoculated with a loop of a 3–4 days old culture from TSA and incubated for 72 h on a horizontal shaker (Novotron, Infors, Switzerland) at 28 °C and 150 rpm. For the main culture 250-ml Erlenmeyer flasks with 100 ml of NB were inoculated with one ml of the pre-culture and incubated at 28 °C and 150 rpm on the horizontal shaker.

2.3. Preparation of samples

The cells were harvested after 24 or 48 h of cultivation upon reaching the stationary growth phase, by centrifugation of the liquid culture for 5 min at 8250g (Biofuge, Heraeus, Germany). The pellet was resuspended with phosphate buffer (pH 7.0), and the centrifugation step was repeated three times. The bacterial suspensions were then adjusted to an optical density of 0.95 ± 0.05 (595 nm) using a UVIKON 922 spectrophotometer (Kontron Instruments, Schliessen, Switzerland). The CPA solutions/suspensions were prepared at different concentrations (Table 4). Albumin, egg yolk, lecithin, soluble starch and skimmed milk were boiled twice in a microwave at 600 W. All other CPAs were autoclaved for 10 min at 115 °C. After cooling to room temperature, suspensions of the micro-organisms were mixed 1:1 with the respective solutions/suspensions of the CPAs. Three ml of the resulting suspension were pipetted in 10-ml serum vials (Ismatec, Germany) and the total weight was determined. The samples were then frozen or freeze-dried in an Advantage EL freeze-dryer (Virtis, Gardiner, USA) using different protocols. At the end of the process the vials were immediately sealed at 0.2 mbar in the freeze-drying chamber of the freeze-dryer. Before determination of the number of viable cells the frozen samples were thawed in a refrigerator. In case of freeze-dried samples sterile, deionised water was added to the original weight.

2.4. Determination of the number of viable cells

The viability of the micro-organisms was determined by the Most Probable Number (MPN) method. 120 µl of NB medium were pipetted in 96-well microtiter plates. After the addition to four wells each of 30 µl of the suspension of micro-organisms, 1:5 dilution steps were made. Based on the number of turbid wells the MPN was calculated with the help of a computer program (Most Probable Number Calculator Version 4.04 1996, Albert J. Klee – Risk Reduction Engineering Laboratory, United States Environmental Protection Agency, Cincinnati, Ohio, USA). Survival rates were calculated by the formula: (viability after treatment/viability before treatment) ×100.

2.5. Comparison of different strains

The four *Pseudomonas* strains I112, Pf153, PCL1391 and CHA0 were prepared as described above. The cells were suspended in skimmed milk as CPA and freeze-dried by cooling from +5 °C to −40 °C in 40 min and drying with a shelf temperature of 5 °C. The MPN was determined before and after freeze-drying.

2.6. Influence of skimmed milk on viability after freezing

Freshly produced micro-organisms of strains Pf153 and PCL1391 were prepared as described above. To this suspension, skimmed milk as CPA or phosphate buffer (control) was added. The samples were frozen from +5 °C to −40 °C in 40 min. The MPN was determined before and after freezing.

2.7. Comparison of different freezing rates

Four different freezing rates were compared for the five strains. Skimmed milk was selected as the CPA. For shock freezing the vials were placed in liquid nitrogen for 2.5 min. The product temperature reached −150 °C. Three other freezing rates were obtained by computer controlled cooling the product temperature from +5 °C to −40 °C over 40, 50 min or 16.7 h. During freezing the product temperature was measured parallel in four different vials. The freezing rate was calculated by the formula: $[T(t + i) − T(t)]/[(t + i) − t]$, with, T = product temperature, t = time, i = time interval. The MPN was determined before and after freezing.

2.8. Comparison of different drying temperatures

The experiments were carried out with strains Pf153 and PCL1391. The drying temperatures of 5, 20 and 30 °C were obtained by adjusting the shelf temperature of the freeze-dryer. The pressure during the process was adjusted to 0.2 mbar. Before drying, all the samples were suspended in skimmed milk and frozen from +5 °C to −40 °C in 40 min. The vials were then divided into three batches which were kept in the freezer at −30 °C until they were dried at the different temperatures. The MPN was determined before and after freeze-drying.

2.9. Effect of different CPAs on the viability after freeze-drying

The influence of 20 different CPAs (Table 4) on the viability of strain Pf153 was compared. The vials were freeze-dried as follows: cooling from +5 °C to −40 °C in 40 min and drying for 18 h with a shelf temperature of 5 °C at 0.2 mbar. The MPN was determined after freeze-drying.

2.10. Influence of the CPAs on storability

Cells of four Pseudomonas strains prepared as described above were suspended in the CPAs saccharose, glucose, lactose, skimmed milk, ligninosulfonic acid or phosphate buffer and freeze-dried (by cooling from +5 °C to −40 °C within 40 min for strains CHA0 and PCL1391, within 50 min for strain I112 and 16.7 h for strain Pf153, and the drying temperature of 30 °C). Afterwards, the vials were stored. The MPN was determined directly after freeze-drying and after 2, 4 and 7 days incubation in a water bath at 40 °C.

2.11. Biocontrol activity of freeze-dried P. putida cells

Strain MF416 was tested against the seed-borne pathogens Alternaria radicina and A. dauci using naturally infected (25% A. dauci, 65% A. radicina) carrot seeds (variety "Laguna" F1). Cells of the same batch were tested as fresh formulated cells (cells mixed with CPAs) and after freeze-drying. The cells were freeze-dried according to the following protocol: CPA: 12.5% lactose, freezing over 16.7 h to −40 °C product temperature, drying over 18 h with shelf temperature of 30 °C and vacuum of 0.2 mbar. Per treatment, 100 seeds were stirred for 15 min in 3 ml suspensions of strain MF416 in 12.5% lactose. Treatments in water or a 0.75% (w/v)

suspension in water of the fungicide Aatiram (active ingredient: 670 g/kg Thiram) were included as controls. The treated seeds were dried at 25 °C for 24 h and sown in domestic polypropylene trays (27.5 × 17.5 × 7 cm) containing 1.5 l pre-moistened horticultural potting substrate (Fruhstorfer Erde Typ P; Archut, Lauterbach, Germany). The seeding rate was 100 seeds per tray and per treatment three trays were sown. The seeds were then covered with approx. 270 ml vermiculite. Afterwards, the trays were loosely covered with a lid and placed in a growth chamber at 20–22 °C and a day/night cycle of 16/8 h. The lids were removed after one week and the plants further cultivated under the same conditions. The number of healthy plants was determined 17–22 days after planting.

2.12. Influence of CPAs on biocontrol activity of P. fluorescens cells

In these experiments cells of the same batch were tested as fresh formulated cells (cells mixed with CPAs) and after freeze-drying. Strain Pf153 was freeze-dried under optimized conditions (freezing over 16.7 h to −40 °C product temperature, drying over 18 h with shelf temperature of 30 °C and vacuum of 0.2 mbar) in the presence of different CPAs (lactose, saccharose or skimmed milk, each 10% in water) and used in biocontrol experiments with *Botrytis cinerea* on detached bean leaves. Six week-old leaves of broad beans (variety "con Amore") were placed in 20 × 20 × 5 cm plexiglass boxes on a stainless steel wire mesh. To achieve constant moisture, 50 ml of deionised water were added to each box. Per treatment, four boxes with three leaves each were used. Using an air brush, one ml of the suspensions was sprayed on each side of the leaves in each box. As control water or 0.5% (w/v) suspension in water of the fungicide Euparen (active ingredient: 50% Tolylfluanid, WG) were included. One milliliter of a 5 × 10⁸ conidial suspension of *B. cinerea* was then sprayed on the upper surface of the leaves and the boxes were incubated at 20 °C and a day/night cycle of 16/8 h. From the second to the fifth day after inoculation the percentage of affected leaf area was assessed and the Area Under Disease Progress Curve (AUDPC) computed.

2.13. Statistical analysis

All experiments were repeated three times time independently. All data were statistically analyzed with the software SAS System for Windows v9.1. All experiments were analyzed with the generalized linear model. Depending on the experimental design for the separation of the means original data, log- or arcsine transformed data were compared with the Student Newman Keuls (SNK) test. Bioassays with replicates in a repetition were analyzed by a block model.

3. Results

3.1. Comparison of different strains

In a first screening, the viability after freeze-drying under not optimized conditions with skimmed milk as CPA was compared

Table 1
Viability (MPN ml⁻¹), before and after freeze-drying, of different Pseudomonas strains.

Strain	Species	Before freeze-drying	Freeze-dried	Survival %
I112	*P. putida*	1.67 (±1.03) × 10⁸	1.60 (±1.56) × 10⁷	10
PCL1391	*P. chlororaphis*	1.44 (±0.59) × 10⁸	3.10 (±2.91) × 10⁶	2
CHA0	*P. protegens*	8.35 (±3.21) × 10⁷	0.84 (±1.25) × 10⁷	10
Pf153	*P. fluorescens*	2.53 (±1.55) × 10⁸	0.94 (±1.02) × 10⁷	4

Data are means and standard deviation of three independent experiments.

Table 2
Influence of different freezing rates, drying temperatures, the occurrence of a cryo-protective agent (CPA) on the viability of the strains Pf153 and PCL1391.

Strain	Freezing rates (°C min⁻¹)	Drying temperature (°C)	CPA	Viability (MPN ml⁻¹ ±SD)[a]		Survival (%)
Pf153	Before freezing			$3.62 (\pm1.24) \times 10^8$	(a)[b]	
	1.3–1.9			$4.45 (\pm7.70) \times 10^6$	(b)	1
	Before freezing		Skimmed milk	$1.35 (\pm0.38) \times 10^8$	(a)	
	1.3–1.9		Skimmed milk	$1.79 (\pm1.59) \times 10^8$	(a)	133
PCL1391	Before freezing			$1.31 (\pm1.26) \times 10^9$	(a)	
	1.3–1.9			$5.92 (\pm8.09) \times 10^7$	(b)	5
	Before freezing		Skimmed milk	$7.75 (\pm2.06) \times 10^8$	(a)	
	1.3–1.9		Skimmed milk	$6.21 (\pm4.14) \times 10^8$	(a)	80
Pf153	Before drying		Skimmed milk	$1.35 (\pm0.38) \times 10^8$	(a)	
	1.3–1.9	5	Skimmed milk	$1.25 (\pm0.42) \times 10^8$	(b)	9
	1.3–1.9	20	Skimmed milk	$6.45 (\pm6.53) \times 10^7$	(ab)	48
	1.3–1.9	30	Skimmed milk	$6.04 (\pm2.99) \times 10^7$	(ab)	45
PCL1391	Before drying		Skimmed milk	$7.75 (\pm2.06) \times 10^8$	(a)	
	1.3–1.9	5	Skimmed milk	$2.63 (\pm1.06) \times 10^7$	(b)	3
	1.3–1.9	20	Skimmed milk	$8.01 (\pm5.65) \times 10^7$	(b)	10
	1.3–1.9	30	Skimmed milk	$2.10 (\pm1.98) \times 10^8$	(b)	27

[a] Each mean and standard deviation are calculated from three independent experiments.
[b] Means of root transformed data of each experiment followed by the same letter are not significantly different following Students-Newman-Keuls test ($P < 0.05$).

Table 3
Influence of different freezing rates on the viability [MPN ml⁻¹] of different pseudomonads formulated with skimmed milk.

Strain	Before freezing	Freezing rates			
		0.6–12 °C/s[b]	1.3–1.9 °C min⁻¹	0.6–0.8 °C min⁻¹	0.04–0.12 °C min⁻¹
I112	1.29×10^9 (a)	7.25×10^8 (a)	9.87×108 (a)	1.90×10^9 (a)	7.26×10^8 (a)
PCL1391	1.15×10^9 (a)	1.03×10^9 (a)	1.02×10^9 (a)	8.48×10^8 (a)	9.19×10^8 (a)
CHA0	6.03×10^8 (a)	4.73×10^8 (a)	1.04×10^9 (a)	6.32×10^8 (a)	8.41×10^8 (a)
Pf153	1.89×10^9 (a)	1.20×10^9 (a)	6.15×10^8 (a)	1.01×10^9 (a)	2.08×10^9 (a)
MF416	1.45×10^9 (a)	8.09×10^8 (a)	8.57×10^8 (a)	7.65×10^8 (a)	9.55×10^8 (a)

[a] Means of three independent experiments. Means of the log₁₀ transformed data within one row followed by the same letter are not significantly different following Student-Newman Keuls test ($P < 0.05$).
[b] Shock freezing with liquid nitrogen.

for four *Pseudomonas* strains. The viability before and after freeze-drying differed dramatically (Table 1). The survival rate ranged between 2% and 10%. None of the tested strains showed acceptable survival rates after freeze-drying.

3.2. Influence of skimmed milk on viability after freezing

For the two strains Pf153 and PCL1391 the viability after freezing was significantly reduced with survival rates lower than 5% when the cells were frozen without the CPA (Table 2). But when skimmed milk was added, the viability before and after freezing did not differ significantly and the survival rates were increased to at least 80%.

3.3. Comparison of different drying temperatures

The different drying temperatures were obtained by variation of the shelf temperature (5, 20 or 30 °C) in the freeze-dryer. The effect of the drying temperature on the viability was tested with two different strains, PCL1391 and Pf153 (Table 2). For strain PCL1391 viability of the cells after drying increased with increasing shelf temperature, although the effect was not significant. Compared to the control (fresh, non freeze-dried cells), the viability of the freeze-dried cells was statistically significant reduced. The survival rate ranged between 3% and 27%. The viability of strain Pf153 did not differ significantly between the freeze-dried and the fresh material at the 20 and 30 °C shelf temperatures. But drying at 5 °C resulted in a significantly lower viability compared to 20 °C and 30 °C.

3.4. Comparison of different freezing rates

Four different freezing rates were compared, of which one was obtained by shock freezing with liquid nitrogen, and three by controlling the shelf temperature in the freeze-dryer. During the freezing process the shelf and/or the product temperature were monitored. With liquid nitrogen the product temperature dropped to −150 °C within 2.5 min. Based on this product temperature a freezing rate of 0.6–12 °Cs⁻¹ was calculated. By computer controlled cooling the shelf of the freeze-dryer a final product temperature of −40 °C was achieved within 40, 50 min or 16.7 h. These cooling processes correspond to freezing rates of 1.3–1.9 °C min⁻¹, 0.6–0.8 °C min⁻¹ or 0.04–0.12 °C min⁻¹, respectively. When the viability of five pseudomonads was determined, for all but one strain the viability was not significantly influenced by the freezing process, irrespective of the freezing rate (Table 3).

3.5. Effect of different CPAs on the viability after freeze-drying

Using strain Pf153, a total of 20 different CPAs was compared. Depending on the CPA used, the MPN value determined after the freeze-drying process differed significantly, varying by six orders of magnitude. Highest viability was achieved by the tested sugars, ligninosulfonic acid and skimmed milk (Table 4). The viability of the cells freeze-dried in the presence of saccharose was 2.9 times higher compared to the cells to which skimmed milk was added. By adding CPAs it was possible to raise the viability by a factor of 440 (Table 4). On the other hand, a number of additives had a negative effect on the viability.

Table 4
Comparison of different cryo-protective agents on the viability of *P. fluorescens*, Pf153 after freeze-drying.

Cryo protective agent	Conc. (% w/v)	Viability (MPN ml^{-1} ±SD) after freeze-drying		IF
Saccharose	12.5	1.02 (±0.44) × 10^9	(a)	439.7
Lactose	12.5	9.11 (±7.70) × 10^8	(a)	392.7
Ligninosulfonic acid	12.5	8.63 (±5.84) × 10^8	(a)	372.0
ᴅ(+)-Glucose-monohydrate	12.5	7.05 (±3.50) × 10^8	(a)	303.9
Skimmed milk 1% fat content	12.5	3.31 (±0.89) × 10^8	(ab)	142.7
Soluble starch	12.5	1.74 (±0.39) × 10^8	(abc)	75.0
Carboxymethylcellulose Na-salt	2.5	1.61 (±0.55) × 10^8	(abc)	69.4
Nutrient broth	12.5	1.27 (±0.35) × 10^8	(abcd)	54.7
Egg white – Albumen	12.5	2.40 (±1.89) × 10^7	(bcde)	10.3
Egg yolk	12.5	2.39 (±1.36) × 10^7	(bcde)	10.3
Lecithin	9.0	3.05 (±4.86) × 10^7	(cde)	13.1
Egg yolk	25.0	2.82 (±3.43) × 10^7	(cde)	12.2
SPAN™ 80/xanthan gum	12.5 / 0.04	2.37 (±2.91) × 10^7	(cde)	10.2
Na-alginate	2.5	1.18 (±0.65) × 10^7	(cde)	5.1
SPAN™ 60/xanthan gum	10.0 / 0.09	1.10 (±0.71) × 10^7	(cde)	4.7
Egg white – Albumen	25.0	9.74 (±11.9) × 10^6	(de)	4.2
Xanthan gum	0.45	4.27 (±5.31) × 10^6	(e)	1.8
Na-glutamate	12.5	4.19 (±4.03) × 10^6	(e)	1.8
Control (without CPA)		2.32 (±1.12) × 10^6	(e)	1.0
Glycerol	12.5	2.66 (±2.01) × 10^4	(f)	0.011
Bentonite	5.0	1.84 (±2.64) × 10^4	(f)	0.008
Activated carbon	5.0	1.29 (±0.30) × 10^2	(g)	0.000
Alkalic lignin	12.5	5.60 (±4.70) × 10^1	(g)	0.000

[a] Each mean and standard deviation are calculated from three independent experiments.
[b] Means of log(10) transformed data followed by the same letter are not significantly different following Students-Newman-Keuls test (*P* < 0.05).
[c] Incrementing factor: viability with CPA/viability of control.

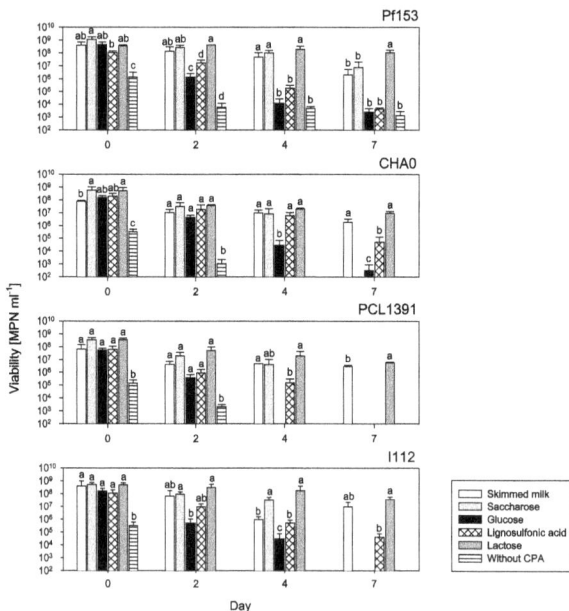

Fig. 1. Viability of freeze-dried cells of strains Pf153, CHA0, PCL1391 and I112 formulated in skimmed milk, saccharose, glucose, ligninosulfonic acid, lactose and without any CPA at 40 °C for up to seven days. Means and SD (error bars) of three independent experiments. Means of the log$_{10}$ transformed data within one time of each isolate followed by the same letter are not significantly different following Student-Newman Keuls test (*P* < 0.05).

3.6. Influence of the CPAs on storability

The five CPAs with best cryoprotective capacity were compared for four different strains (Fig. 1). Compared to freeze-drying

without any additive, survival of all strains was improved in the presence of the selected CPAs. Storability at 40 °C of all strains was differently affected by the CPAs. For all strains glucose and ligninosulfonic acid protected the cells during freeze-drying, but

Table 5
Viability [MPN ml^{-1}] before and after freeze-drying of *P. putida* MF416 and *P. fluorescens* Pf153.

Strain	CPA	Before freeze-drying	After freeze-drying
MF416	Lactose	7.35 (±1.26) × 10^8 (a)	2.86 (±1.49) × 10^8 (b)
Pf153	Lactose	1.15 (±0.28) × 10^9 (a)	2.52 (±2.72) × 10^9 (a)
Pf153	Saccharose	1.87 (±0.40) × 10^9 (a)	1.55 (±1.62) × 10^9 (a)
Pf153	Skimmed milk	1.87 (±0.40) × 10^9 (a)	1.75 (±1.20) × 10^9 (a)

Means and SD of three independent experiments. Means of the data within one row followed by the same letter are not significantly different following Student-Newman Keuls test ($P < 0.05$).

Table 6
Efficacy of freeze-dried and not freeze-dried cells of *P. putida* MF416 against *A. radicina* and *A. dauci* on carrot seeds.

Treatment	Number of healthy plants (%)
Water control	16.3 ± 5.5 (c)
Fresh cells	35.1 ± 2.2 (b)
Freeze-dried cells	35.3 ± 2.9 (b)
Chemical control (Thiram)	72.2 ± 8.3 (a)

Means and SD of three independent experiments. Means of the data with same letter are not significantly different following Student-Newman Keuls test ($P < 0.05$).

the viability declined dramatically during storage. Lactose followed by skimmed milk, was for all strains the best CPA. Saccharose protected the cells of all strains over four days storage but after seven days only for strain Pf153 saccharose gave comparable results to skimmed milk (Fig. 1).

3.7. Bioassays

To evaluate the influence of freeze-drying on the efficacy of pseudomonads against phytopathogens, two different pathosystems were used. In the first experiment the biocontrol activity of fresh and freeze-dried cells was compared. Strain Mf416 was tested for the ability to control the seed-borne diseases leaf blight and black rot of carrots caused by *A. dauci* and *A. radicina*, respectively. The efficacy of the freeze-dried cells was nearly identical to that of the fresh cells (Table 6) even though the viability declined after freeze-drying (Table 5). The efficacy of the freeze-dried or fresh cells of strain Mf416, was however significantly lower than the chemical standard (Table 6).

In the second experiment the influence of the CPAs on the biocontrol activity was evaluated. Strain Pf153 was tested against gray mould caused by *B. cinerea* on broad bean leaves. Viability of strain Pf153 was not reduced after freeze-drying, irrespective of the chosen CPA (Table 5). In contrast, the CPAs had a significant influence on the disease controlling capability of strain Pf153. (Table 7).

Table 7
Influence of different CPAs on the efficacy of *P. fluorescens* Pf153 against *B. cinerea* on detached leaves of broad beans.

Treatment	AUDPC before freeze-drying	AUDPC after freeze-drying
Water control	249 ± 69 (b)	234 ± 70 (b)
Lactose	174 ± 50 (c)	184 ± 20 (c)
Saccharose	253 ± 28 (b)	295 ± 22 (a)
Skimmed milk	146 ± 51 (c)	169 ± 37 (c)
Chemical control (Euparen)	56 ± 29 (d)	66 ± 30 (d)

Means ± SD of three independent experiments. Means of the data within one column followed by the same letter are not significantly different following Student-Newman Keuls test ($P < 0.05$). (AUDPC = Area under disease progress curve).

Fresh and freeze-dried cells formulated in skimmed milk or lactose were equally effective, whereas no significant control was seen when strain Pf153 was formulated in saccharose.

4. Discussion

Freeze-drying is the preferred drying technique for the conservation of micro-organisms (Morgan et al., 2006). It has been reported by several authors (Carvalho et al., 2004b; Heckly, 1985; Montel Mendoza et al., 2014) that micro-organisms belonging to different species and strains may differ in their sensitivity to freeze-drying. In the present study we investigated the potential of freeze-drying as a method to formulate bacteria of the genus *Pseudomonas* for use as biocontrol agents. Using a not optimized freeze-drying protocol, survival rates of only 2–10% for four different *Pseudomonas* species correspond to results of Palmfeldt et al. (2003) achieving for *P. chlororaphis* survival rates of 0.6–26%.

Because of these low survival rates, we studied the freeze-drying process in order to identify critical factors influencing the viability. Firstly, the initial step of freezing of the cells was analyzed. The results demonstrated that freezing without CPAs dramatically affected the survival of the two strains of *Pseudomonas* tested. The observed positive effect of the CPA skimmed milk on the survival of the two strains after freezing confirms the results of previous studies (Berny and Hennebert, 1991; Engel, 1992; Gehrke, 1991; Palmfeldt et al., 2003; Tsvetkov and Brankova, 1983). King and Su (1993) explained the positive effect of skimmed milk on the survival of the freezing by the formation of a thin layer of milk protein over the cell wall proteins. Additionally, calcium ions from the skimmed milk may contribute to the protective effect (King and Su, 1993).

Because the freezing phase has been described as the most critical step within the freeze-drying process (Gehrke, 1991), we investigated several freezing rates using 12.5% skimmed milk as CPA. Although the freezing process has been discussed in several publications (Clement, 1961; Engel, 1992; Mazur, 1970), only incomplete data are available on the temperature changes in the product temperature during this step. Our data demonstrate that, especially for high freezing rates, the shelf temperature does not sufficiently represent the product temperature (data not shown). Therefore, realistic freezing rates can only be calculated on the basis of product temperature. Abadias et al. (2001b) suggested that freezing with liquid nitrogen is probably too fast to allow the internal water to migrate out of the cell and that water frozen inside the cell leads to internal damage. Nei (1973) found that freezing of *Saccharomyces cerevisiae* and *Escherichia coli* at fast freezing rates affected the cells by creating ice nucleates. Gehrke (1991) concluded that the optimal freezing rate depends on the ability of the cell to release the water. Consequently, freezing rates for large cells with thick cell walls are lower than for small cells. Berny and Hennebert (1991) suggested that the freezing rate is most safe when the cells do not loose water during the freezing process and reach the eutectic point in a frozen, amorph stage. Our results demonstrate that the freezing rate is only slightly influencing the viability after freezing.

According to Gehrke (1991) the cells survive the sublimation or drying phase without cell damage. As long as the free water and not the bounded water is sublimated, the cells are not damaged. Our results demonstrate that during sublimation the viability was influenced by the drying temperature. Higher drying temperatures resulted in a higher viability after freeze-drying. One explanation could be, that possibly the viability was affected by sublimation of bounded water at lower drying temperatures. Additionally, depending on the CPA the drying temperature was influencing the product texture. In our experiments with 30 °C shelf temperature lactose was melted, the product structure collapsed with the effect that the solubility after freeze-drying was reduced.

Cpas play an important role in the freeze-drying process (Morgan et al., 2006). The CPAs have two main functions: they protect the living cells biochemically against damage during freeze-drying and provide a dry residue with defined physical structure acting as a support material and as receptor in rehydration (Berny and Hennebert, 1991). When strain Pf153 was suspended in saccharose, lactose, ligninosulfonic acid, glucose or skimmed milk, the viability increased to two decimal powers compared to cells without protection. Comparable results of a protective effect of lactose were found by Cabrefiga et al. (2014). The unfrozen water (UFW) of the CPA may influence the survival after freeze-drying (Oetjen and Haseley, 2004). Although the increasing UFW of egg white (6%), yolk (13%), glucose (29%), sucrose (36%) and lactose (41%) correspond to the increasing viability of strain Pf153, glycerol with an UFW of 46% did not protect the cells of strain Pf153. Therefore, it is unlikely that the percentage of UFW in the CPA is the only important factor influencing the survival after freeze-drying. Disaccharides like saccharose or lactose are acting as a CPA, because of its ability to hydrate biological structures like proteins and membranes (Arakawa et al., 2001; Crowe et al., 2001). This effect is also called as water replacement hypothesis. Tsvetkov and Brankova (1983) reported when a Micrococcus strain was suspended in 5% saccharose, the viability increased to 79% compared to unformulated cells. Abadias et al. (2001a) obtained no significant differences in the viability when Candida sake was freeze-dried with saccharose or lactose. But better survival rates of 22% were achieved with skimmed milk (Abadias et al., 2001a) as CPA. Zayed and Roos (2004) obtained comparable results when Lactobacillus salivarius was freeze-dried with skimmed milk or saccharose. In skimmed milk, lactose is the dominant sugar component. Therefore, possibly lactose is the relevant cryo-protective substance of skimmed milk. Otherwise, Palmfeldt et al. (2003) and Walsh et al. (2000) have shown that a combination of different CPAs enhance the viability after freeze-drying. Therefore, synergistic effects of the different ingredients of skimmed milk cannot be excluded.

Our results confirm that skimmed milk and several sugars are suitable CPAs for pseudomonads. Additionally, CPAs like ligninosulfonic acid are interesting substances for freeze-drying because they have some other characteristics which are important for the formulation of biocontrol agents. Ligninosulfonic acid can be used as UV-protectant and has also some effects on phytopathogens (unpublished data). We also confirmed that some substances cannot protect cells like it was described for bentonite by Stephan and Zimmermann (1998) and for glycerol by Abadias et al. (2001a).

After optimizing the different parameters (freezing rate, drying temperature, CPA) the viability of strain Pf153 after freeze-drying was not significantly reduced. This is one of the first reports that a technology is described with which desiccation sensitive pseudomonads can be dried without loss of viability.

A good storability of living cells is a prerequisite for the commercialization of a biocontrol agent (Burges, 1998). Therefore, in the next set of experiments we investigated the influence of the five best CPAs on the survival of four different Pseudomonas strains during storage. We confirmed the results of other authors (Sidyakina and Lozitskaya, 1991; Stephan and Zimmermann, 2001; Zayed and Roos, 2004), that skimmed milk is a sufficient protectant to enhance the storability of micro-organisms. Anyhow, within our storage experiments at 40 °C all tested strains survived slightly better when cells were formulated with lactose than with skimmed milk. Saccharose was only suitable for strain Pf153 after storage for seven days at 40 °C but not for the other three strains. Palmfeldt et al. (2003) demonstrated that saccharose is suitable for

freeze-drying of P. chlororaphis in concentrations between 5% and 13%. They verified that the intra- and extra-cellular saccharose concentration was nearly similar and refer it to the incorporation of saccharose in the cytoplasmatic membrane, before the cells were frozen. Additionally, saccharose is absorbed as a disaccharide and is not hydrolyzed to the monosaccharide form (Palmfeldt et al., 2003). Therefore, one explanation of our results could be that the strains CHA0, PCL1391 and I112 are able to hydrolyze saccharose before freezing with the consequence of lower protection during storage or the incorporation of saccharose in the cytoplasmatic membrane differed.

Independently of the selected CPAs the Pseudomonas strains differed in their survival after storage. When no CPAs were added after 2 days storage no viable cells were found for the strain I112. In contrast, for strain Pf153 some cells were viable even after 7 days storage at 40 °C. Differences in the storability of micro-organisms can be caused by fermentation conditions (Carvalho et al., 2003, 2004b; Hallsworth and Magan, 1994; Jackson et al., 1997; Zhang et al., 2006). Within our experiments all pseudomonads were produced under same conditions and the fermentation was not optimized for each isolate. Therefore, it cannot be excluded that the differences in the storability of the isolates will change after optimizing fermentation parameters.

It is well described that pseudomonads are effective antagonists of phytopathogens (Fuchs et al., 2000; Maurhofer et al., 1992; Meena et al., 2002; Vidhyasekaran and Muthamilan, 1999). But often, within the formulation process, tested additives are toxic or based on vigorous physical manipulations the viability and/or efficacy of potential antagonists is reduced (Bowers, 1982). Therefore, in the last set of experiments we compared the efficacy of freshly produced and freeze-dried cells in two different ad planta bioassay systems.

In both systems the efficacy of the freeze-dried cells was as good as the freshly produced cells. Therefore, we were able to demonstrate that freeze-dried pseudomonads can control seed-borne and plant diseases as good as freshly produced cells. But the efficacy of the pseudomonads was not as good as the chemical control. Therefore, further investigations have to be carried out to enhance the efficacy by optimizing the formulations and application technology.

Additives for the formulation of microbial biopesticides have various functions (Burges, 1998). They can act e.g. as carriers or fillers, they can affect the physical properties of formulations, or they can have synergistic effects. Therefore, we compared as well the influence of three different CPAs on the efficacy of freeze-dried cells of strain Pf153 against B. cinerea. Although the viability before and after freeze-drying was not significantly different, strain Pf153 formulated in saccharose was not as effective as formulated in lactose or skimmed milk. One possibility is that the CPAs can be utilized by the antagonist or by the pathogen. Elmer and Reglinski (2006) summarized that biological suppression of B. cinerea arises via competition of nutrients and space, the production of inhibitory metabolites or parasitism and can stimulate defense mechanisms. Therefore, in additional experiments we compared the growth of the antagonist and the pathogen on agar or in liquid culture by adding the CPAs as carbohydrate source (data not shown). For the pathogen best growth was achieved by adding lactose and skimmed milk, followed by glucose and saccharose. When the sporulation of B. cinerea on the different media was investigated only on the medium containing skimmed milk B. cinerea produced conidia. On all other media B. cinerea did not sporulate in the investigated time. On the other hand for the antagonistic pseudomonad strain Pf153 no preference to a specific carbohydrate source was seen. Therefore, it is unlikely, that the differences in the efficacy can be explained in the way that only specific CPAs can be utilized by the two micro-organisms on the leaf surface. Possibly, CPAs influence the physical properties of the spray deposit, and therefore its stability on the leaf surface.

Our results demonstrate that after optimizing the freeze-drying process pseudomonads can be freeze-dried without loss of viability. The results underline that particularly by selecting the right CPA beside viability, storability and as well efficacy of pseudomonads can be enhanced.

Acknowledgments

This research was supported by SafeCrop Centre, funded by Fondo per la Ricerca, Autonomous Province of Trento. We have to thank Ben Lugtenberg (University Leiden, The Netherlands), Monika Maurhofer (ETH, Switzerland), Eckhard Koch (JKI, Germany) for providing us with *Pseudomonas* strains.

References

Abadias, M., Benabarre, A., Teixido, N., Usall, J., Vinas, I., 2001a. Effect of freeze drying and protectants on viability of the biocontrol yeast *Candida sake*. Int. J. Food Microbiol. 65, 173–182.

Abadias, M., Teixido, N., Usall, J., Vinas, I., Magan, N., 2001b. Improving water stress tolerance of the biocontrol yeast *Candida sake* grown in molasses-based media by physiological manipulation. Can. J. Microbiol. 47, 123–129.

Arakawa, T., Prestrelski, S.J., Kenney, W.C., Carpenter, J.F., 2001. Factors affecting short-term and long-term stabilities of proteins. Adv. Drug Deliv. Rev. 46, 307–326.

Berny, J.F., Hennebert, G.L., 1991. Viability and stability of yeast-cells and filamentous fungus spores during freeze-drying – effects of protectants and cooling rates. Mycologia 83, 805–815.

Bora, T., Ozaktan, H., Gore, E., Aslan, E., 2004. Biological control of *Fusarium oxysporum* f. sp *melonis* by wettable powder formulations of the two strains of *Pseudomonas putida*. J. Phytopathol. 152, 471–475.

Bowers, R.C., 1982. Commercialization of microbial biological control agents. In: Charudattan, R., Walker, H.L. (Eds.), Biological Control of Weeds with Plant Pathogens. John Wiley & Sons, New York, pp. 157–173.

Burges, H.D., 1998. Formulation of Microbial Biopesticides. Kluwer Academic Publishers, Dordrecht, Boston, London.

Cabrefiga, J., Frances, J., Montesinos, E., Bonaterra, A., 2014. Improvement of a dry formulation of *Pseudomonas fluorescens* EPS62e for fire blight disease biocontrol by combination of culture osmoadaptation with a freeze-drying lyoprotectant. J. Appl. Microbiol. 117, 1122–1131.

Carvalho, A.S., Silva, J., Ho, P., Teixeira, P., Malcata, F.X., Gibbs, P., 2003. Effect of various growth media upon survival during storage of freeze-dried *Enterococcus faecalis* and *Oenococcus oeni durans*. J. Appl. Microbiol. 94, 947–952.

Carvalho, A.S., Silva, J., Ho, P., Teixeira, P., Malcata, F.X., Gibbs, P., 2004a. Effects of various sugars added to growth and drying media upon thermotolerance and survival throughout storage of freeze-dried *Lactobacillus delbrueckii* ssp *bulgaricus*. Biotechnol. Prog. 20, 248–254.

Carvalho, A.S., Silva, J., Ho, P., Teixeira, P., Malcata, F.X., Gibbs, P., 2004b. Relevant factors for the preparation of freeze-dried lactic acid bacteria. Int. Dairy J. 14, 835–847.

Clement, M.T., 1961. Effects of freezing, freeze-drying, and storage in frozen and frozen state on viability of *Escherichia coli* cells. Can. J. Microbiol. 7, 99–8.

Commare, R.R., Nandakumar, R., Kandan, A., Suresh, S., Bharathi, M., Raguchander, T., Samiyappan, R., 2002. *Pseudomonas fluorescens* based bio-formulation for the management of sheath blight disease and leaffolder insect in rice. Crop Prot. 21, 671–677.

Commission, E., 2015.

Crowe, J.H., Crowe, L.M., Oliver, A.E., Tsvetkova, N., Wolkers, W., Tablin, F., 2001. The trehalose myth revisited: Introduction to a symposium on stabilization of cells in the dry state. Cryobiology 43, 89–105.

Dandinrand, L.M., Morra, M.J., Chaverra, M.H., Orser, C.S., 1994. Survival of *Pseudomonas* spp in air-dried mineral powders. Soil Biol. Biochem. 26, 1423–1430.

Elmer, P.A.G., Reglinski, T., 2006. Biosuppression of *Botrytis cinerea* in grapes. Plant. Pathol. 55, 155–177.

Engel, G., 1992. The influence of freezing and storage in liquid-nitrogen on the survivability of yeast-cells and mold spores. Kieler milchwirtschaftliche Forschungsberichte 44, 211–216.

Fang, G.C., Waldrup, V.C., Wechter, W.P., Kluepfel, D.A., 2007. A broad-spectrum antagonistic activity of the biocontrol agent *Pseudomonas synxantha* BG33R. Phytopathology 97, S34–S34.

Fuchs, J.G., Moenne-Loccoz, Y., Defago, G., 2000. The laboratory medium used to grow biocontrol *Pseudomonas* sp Pf153 influences its subsequent ability to protect cucumber from black root rot. Soil Biol. Biochem. 32, 421–424.

Gehrke, H.H., 1991. Untersuchungen zur Gefriertrocknung von Mikroorganismen. VDI Verlag, Düsseldorf.

Hallsworth, J.E., Magan, N., 1994. Improved biological control by changing polyols/ trehalose in conidia of entomopathogens. In: Brighton Crop Protection Conference, 1091–1096.

Heckly, R.J., 1985. Principles of preserving bacteria by freeze-drying. Dev. Ind. Microbiol. 26, 379–395.

Hirano, S.S., Upper, C.D., 2000. Bacteria in the leaf ecosystem with emphasis on *Pseudomonas syringae* – a pathogen, ice nucleus, and epiphyte. Microbiol. Mol. Biol. Rev. 64, 624.

Jackson, M.A., McQuire, M.R., Lacey, L., Wraight, S., 1997. Liquid culture production of desiccation tolerant blastospores of the bioinsecticidal fungus *Paecilomyces fumosoroseus*. Mycol. Res. 101, 35–41.

Johansson, P.M., Wright, S.A.I., 2003. Low-temperature isolation of disease-suppressive bacteria and characterization of a distinctive group of pseudomonads. Appl. Environ. Microbiol. 69, 6464–6474.

King, V.A.E., Su, J.T., 1993. Dehydration of *Lactobacillus acidophilus*. Process Biochem. 28, 47–52.

Krimm, U., Banda-Nkpwatt, D., Schwab, W., Schreiber, L., 2005. Epiphytic microorganisms on strawberry plants (*Fragaria ananassa* cv. Elsanta): identification of bacterial isolates and analysis of their interaction with leaf surfaces. FEMS Microbiol. Ecol. 53, 483–492.

Mathivanan, N., Prabavathy, V.R., Vijayanandraj, V.R., 2005. Application of talc formulations of *Pseudomonas fluorescens* nigulo and *Trichoderma viride* pers. Ex SF Gray decrease the sheath blight disease and enhance the plant growth and yield in rice. J. Phytopathol. 153, 697–701.

Maurhofer, M., Keel, C., Schnider, U., Voisard, C., Haas, D., Defago, G., 1992. Influence of enhanced antibiotic production in *Pseudomonas fluorescens* strain-Cha0 on its disease suppressive capacity. Phytopathology 82, 190–195.

Mazur, P., 1970. Cryobiology: freezing of biological systems–the responses of living cells to ice formation are of theoretical interest and practical concern. Science 168, 939–949.

Meena, B., Radhajeyalakshmi, R., Marimuthu, T., Vidhyasekaran, P., Velazhahan, R., 2002. Biological control of groundnut late leaf spot and rust by seed and foliar applications of a powder formulation of *Pseudomonas fluorescens*. Biocontrol Sci. Technol. 12, 195–204.

Moenne-Loccoz, Y., Naughton, M., Higgins, P., Powell, J., O'Connor, B., O'Gara, F., 1999. Effect of inoculum preparation and formulation on survival and biocontrol efficacy of *Pseudomonas fluorescens* F113. J. Appl. Microbiol. 86, 108, 87, 787–787.

Montel Mendoza, G., Pasteris, S.E., Otero, M.C., Nader-Macias, M.E.F., 2014. Survival and beneficial properties of lactic acid bacteria from raniculture subjected to freeze-drying and storage. J. Appl. Microbiol. 116, 157–166.

Morgan, C.A., Herman, N., White, P.A., Vesey, G., 2006. Preservation of micro-organisms by drying; a review. J. Microbiol. Methods 66, 183–193.

Nei, T., 1973. Some aspects of freezing and drying of microorganisms on basis of cellular water. Cryobiology 10, 403–408.

O'Callaghan, M., Swaminathan, J., Lottmann, J., Wright, D., Jackson, T., 2006. Seed coating with biocontrol strain *Pseudomonas fluorescens* F113. N. Z. Plant Prot. 59.

Oetjen, G.-W., Haseley, P., 2004. Freeze-drying. Wiley-VCH, Weinheim.

Palmfeldt, J., Radstrom, P., Hahn-Hagerdal, B., 2003. Optimisation of initial cell concentration enhances freeze-drying tolerance of *Pseudomonas chlororaphis*. Cryobiology 47, 21–29.

Rajappan, K., Vidhyasekaran, P., Sethuraman, K., Baskaran, T.L., 2002. Development of powder and capsule formulations of *Pseudomonas fluorescens* strain Pf-1 for control of banana wilt. Zeitschrift fur Pflanzenkrankheiten und Pflanzenschutz J. Plant Dis. Prot. 109, 80–87.

Senthil, N., Raguchander, T., Viswanathan, R., Samiyappan, R., 2003. Talc formulated fluorescent pseudomonads for sugarcane red rot suppression and enhanced yield under field conditions. Sugar Tech. 5, 37–43.

Sidyakina, T.M., Lozitskaya, N.D., 1991. Viability of bacteria *Pseudomonas dentrificans* Bkm B-892 subjected to lyophilization and cryopreservation in different protective media. Kriobiologiya 1, 33–39.

Someya, N., Tsuchiya, K., Yoshida, T., Noguchi, M.T., Sawada, H., 2007. Encapsulation of cabbage seeds in alginate polymer containing the biocontrol bacterium *Pseudomonas fluorescens* strain LRB3W1 for the control of cabbage soilborne diseases. Seed Sci. Technol. 35, 371–379.

Stephan, D., Zimmermann, G., 1998. Development of a spray-drying technique for submerged spores of entomopathogenic fungi. Biocontrol Sci. Technol.

Stephan, D., Zimmermann, G., 2001. Locust control with *Metarhizium flavoviride*: drying and formulation of submerged spores. In: Koch, E., Leinonen, P. (Eds.), Cost Action 830 Microbial Inoculants for Agriculture and Environment – Formulation of Microbial Inoculants, European Commission, EUR 19692, pp. 27–34.

Tsvetkov, T., Brankova, R., 1983. Viability of micrococci and lactobacilli upon freezing and freeze-drying in the presence of different cryoprotectants. Cryobiology 20, 318–323.

Vidhyasekaran, P., Muthamilan, M., 1999. Evaluation of a powder formulation of *Pseudomonas fluorescens* Pf1 for control of rice sheath blight. Biocontrol Sci. Technol. 9, 67–74.

Walsh, U., Morrissey, J., O'Gara, F., 2000. *Pseudomonas* for biocontrol of phytopathogens: from molecular genomics to commercial exploitation. Curr. Opin. Biotechnol. 12, 289–295.

Weller, D.M., 2007. *Pseudomonas* biocontrol agents of soilborne pathogens: looking back over 30 years. Phytopathology 97, 250–256.

Zayed, G., 2004. Influence of trehalose and moisture content on survival of *Lactobacillus salivarius* subjected to freeze-drying and storage. Process Biochem. 39, 1081–1086.

Zhang, S.A., Schulte, D.A., Jackson, M.A., Boehm, M.J., Slininger, P.J., Liu, Z.L., 2006. Cold shock during liquid production increases storage shelf-life of *Cryptococcus nodaensis* OH 182.9 after air-drying. Biocontrol Sci. Technol. 16, 281–293.

Influence of different growth conditions on the survival and the efficacy of freeze-dried *Pseudomonas fluorescens* strain Pf153 (2015).

Bisutti I.L., Hirt K. and Stephan D.

Biocontrol Science and Technology 25(11): 1269-1284

http://dx.doi.org/10.1080/09583157.2015.1044498

Biocontrol Science and Technology, 2015
Vol. 25, No. 11, 1269–1284, http://dx.doi.org/10.1080/09583157.2015.1044498

Taylor & Francis
Taylor & Francis Group

RESEARCH ARTICLE

Influence of different growth conditions on the survival and the efficacy of freeze-dried *Pseudomonas fluorescens* strain Pf153

I.L. Bisutti, K. Hirt and D. Stephan*

Julius Kühn-Institut, Federal Research Centre for Cultivated Plants, Institute for Biological Control, Darmstadt, Germany

(*Received 27 November 2014; returned 3 January 2015; accepted 18 April 2015*)

High viability, storability and tolerance to variable environmental conditions are key factors in the development of microbial biological control agents (BCAs). The efficacy of microbial BCAs is influenced by the culture conditions and formulation process. Therefore, we investigated the influence of diverse growth conditions on the survival during freeze-drying and on the biocontrol efficacy of *Pseudomonas fluorescens* strain Pf153. Culture time, temperature and media, mild heat shock and pH change influenced the bacterium viability after freeze-drying. The best survival rate was reached by cultivation in King's broth for 16 or 20 h. Growth temperatures of 25 and 30°C and a mild heat shock at 35°C for one hour influenced the survival rate positively. In all bioassays against *Botrytis cinerea* on *Vicia faba* leaves, Pf153 showed a significant increased efficacy compared to the untreated control. No differences of the efficacy between fresh and freeze-dried cells were observed. Furthermore, a growth temperature of 20°C increased the efficacy of Pf153 against *B. cinerea*. These results underline that the quality of the formulated product can be improved by manipulating the fermentation process.

Keywords: harvesting time; liquid fermentation; lyophilisation; media; temperature; pseudomonades

1. Introduction

Biological control agents (BCAs) based on micro-organism are an important component of integrated pest management for control of plant pathogens and insect pests (Radja Commare et al., 2002; Spadaro & Gullino, 2005). The application of micro-organisms as biofertilisers and BCAs has become increasingly important, not only to improve plant growth and to manage plant diseases and pests, but also to avoid environmental pollution (Mathivanan, Prabavathy, & Vijayanandraj, 2005). Additionally, BCAs have received increasing levels of attention over the past two decades because of consumers' concerns regarding the residues of chemical pesticides (Liu et al., 2014).

A successful application of microbial BCAs depends on the development of a suitable and economically feasible formulation. Formulations must deliver a final product with optimised efficacy (Bora, Ozaktan, Gore, & Aslan, 2004), a user friendly application method and with long storage stability (Costa, Usall, Teixido, Torres, & Vinas, 2002; Selvaraj et al., 2014). In fact, the formulation process is very

---

*Corresponding author. Email: dietrich.stephan@jki.bund.de

important during BCA development, when they are transformed from laboratory tested to field applied products (Liu et al., 2014). Formulations can contain spore-forming microbes (e.g. sporulating fungi and Gram-positive bacteria), or metaboli-cally active micro-organism (e.g. Gram-negative bacteria) and are present in different physical forms. Generally, formulations are either in liquid, powder or granular form, and contain additives that can improve the physical properties of the formulation, or can be utilised by the micro-organisms when applied. BCAs are often less consistent in their performance than chemicals, but often this inconsistency is caused by inappropriate formulations (Paau, 1988). Indeed, the efficacy is strongly dependent on the formulation, and the formulation depends on the fermentation type (Abadias, Benabarre, Teixidó, Usall, & Viñas, 2001; Spadaro & Gullino, 2005).

Freeze-drying is known to be a most convenient and successful method of preserving bacteria, yeasts and sporulating fungi (Berny & Hennebert, 1991). This method maintains the best viability but is very expensive. At an industrial level, other drying methods (e.g. spray-drying) are used, but some micro-organisms, like Gram-negative bacteria, lose their viability within the spray-drying process (Montesinos, 2003). Freeze-drying assures long viability, storability, protection from contamination during storage, and ease of product distribution for different micro-organisms (Miyamoto-Shinohara et al., 2000; Smith & Onions, 1983). Therefore, this drying technique is commonly used to preserve bacterial cultures in research and industry (Morgan, Herman, White, & Vesey, 2006; Palmfeldt, Rådström, & Hahn-Hägerdal, 2003). Although highly successful, this process is complex and not suitable for all micro-organisms, as genetic damage can occur by over-drying (Smith & Onions, 1983). Unfortunately, this conservation method can also decrease cell viability due to the extreme process conditions (Palmfeldt & Hahn-Hägerdal, 2000). During this treatment, the bacterial cells are exposed to the process of freezing and drying: the cells are subjected to a number of stresses, including high concentration of solutes, extreme pH, low temperature, formation of ice crystals and sublimation of water out of the cell (Zhao & Zhang, 2005). It is known that the survival rate is dependent on the micro-organism genus. Moreover, the survival rate after freeze-drying is higher for Gram-positive (i.e. genus *Brevibacterium* and *Corynebacterium*) than for Gram-negative bacteria (i.e. genus *Pseudomonas*), 80% and 50% respect-ively. For yeast, survival rates of 10% or less have been reported (Miyamoto-Shinohara et al., 2000). Bacterial survival during freeze-drying is not only influenced by the bacterial species but also by external factors, such as the suspending medium used, harvesting time, growing conditions like fermentation medium cell concentra-tions, or the freeze-drying and rehydrating conditions (Heckly, 1985).

Fluorescent pseudomonas, particularly *Pseudomonas fluorescens* Migula, are the bacterial antagonists considered one of the best candidates for biocontrol (Mathivanan et al., 2005). They are known to improve plant growth and to exert plant disease control in a variety of crops (Radja Commare et al., 2002; Mathivanan et al., 2005). The growth of plant pathogens is inhibited by diverse mechanisms, one example being the production of antibiotics and siderophores (Mathivanan et al., 2005). *P. fluorescens* has been shown to be effective against different diseases when tested under field and greenhouse conditions. It has displayed efficacy against *Fusarium udum* on pigeon pea (Vidhyasekaran, Sethuraman, Rajappan, & Vasumathi, 1997), late leaf spot of groundnut (Meena, 2010) and *Pyricularia oryzae* (blast disease) in rice (Vidhyasekaran, Rabindran, et al., 1997). *P. fluorescens* reduced disease intensity of sheath blight caused by *Rhizoctonia solani* (Kuhn) (Mathivanan et al., 2005; Nandakumar, Babu, Viswanathan, Raguchander, & Samiyappan, 2001; Rabindran & Vidhyasekaran, 1996; Radja Commare et al., 2002), and also leaf folder incidence in rice (Radja Commare et al., 2002) or downy mildew in pearl millet (Umesha, Dharmesh, Shetty, Krishnappa, & Shetty, 1998). In addition, it displayed growth promotion activity and yield increase in tomato (Manikandan, Saravanakumar, Rajendran, Raguchander, & Samiyappan, 2010) and banana crops (Selvaraj et al., 2014), and was also effective against *Fusarium* wilt in both crops. In particular, *P. fluorescens* strain Pf153 reduced disease symptoms in greenhouse and field trials, and protected cucumber against a number of pathogens (*Phomopsis sclerotioides*, *Fusarium oxysporum* and *Pythium ultimum*) in the gnotobiotic system (Fuchs & Defago, 1991). Isolated from roots of tobacco grown in Morens soil (Canton Fribourg, Switzerland), which is suppressive to *Thielaviopsis basicola*, Pf153 synthesised different antifungal compounds, including an extracellular protease and hydrogen cyanide. Unlike other pseudomonads from Morens soil, it neither produces 2,4-diacetylphloroglucinol nor pyoluteorin (Fuchs, Moënne-Loccoz, & Défago, 2000; Fuchs et al., 2000). It was highly antagonistic in *in vitro* tests, against *Aspergillus flavus* (Thakur & Rao, 2001), and it reduced kernel infection of groundnut when used as seed dressing and as soil application (Thakur, Rao, & Subramanyam, 2003). It also inhibited *Phomopsis sclerotioides* on malt agar, potato dextrose agar and King's medium B agar (Fuchs & Defago, 1991).

Because of its biocontrol potential this *Pseudomonas* strain was chosen to investigate the influence of different growth conditions on cell viability after freeze-drying and on their efficacy against *Botrytis cinerea* on *Vicia faba* leaves. In particular the effect of harvesting time, growth media, temperature, pH changes and mild heat shock on the survival of the bacterium after the conservation process were studied. The influence of freeze-drying, growth media, temperature and mild heat shock on the efficacy was evaluated.

2. Materials and methods

2.1. Cultivation of P. fluorescens Pf153

The bacterial *P. fluorescens* strain Pf153, provided by the ETH Zürich (Switzerland), was routinely cultivated on Tryptic Soy Agar (TSA 30 g Tryptic Soy Broth (TSB, Difco, Germany) and 15 g agar-agar (Roth, Germany) in 1000 ml de-ionised water) at 25°C. The pre-culture was prepared in 30 ml of autoclaved King's medium B (KB) (King, Ward, & Raney, 1954) in a 50-ml Erlenmeyer flask. The flask was then incubated for 24 h on a horizontal shaker (Novotron, Infors, Switzerland) at 28°C and 150 rpm.

2.2. Cultivation and preparation

The experiments were performed in 100-ml Erlenmeyer flasks or in a 4-l fermenter (ISF-100, Infors). The flasks were filled with 30 ml, while the fermenter with 3 l KB medium (20.0 g proteose peptone No. 3 (Difco), 10 ml glycerol (Merck, Germany), 1.5 g KH_2PO_4 (Merck), 1.5 $MgSO_4 \times 7 H_2O$ (Merck) unless otherwise specified) and autoclaved at 121°C for 20 min. The inoculation was carried out at a ratio of 1:100. The flasks were then incubated on a horizontal shaker at 150 rpm. The fermenter was kept at 28°C, 900 rpm agitation and 3.70 Nl min^{-1} air. Dissolved oxygen and pH were monitored throughout fermentation. To avoid foam formation, Antifoam (Roth) was added in automatic mode.

After fermentation, Pfl53 cells were harvested by centrifugation (5 min at 8000 g) (Biofuge, Heraeus, Germany) and washed three times with phosphate buffer (12.5 mM, pH 7.0). The resulting pellet was re-suspended in phosphate buffer to obtain an optical density (OD) at 595-nm of 0.95 (±0.02) using a UVIKON 922 spectrophotometer (Kontron Instruments, Switzerland). Like described by Stephan, Bisutti, and Matos da Silva (2007) before freeze-drying, a 20% (w/v) solution of lactose (Edelweiss, Germany) was added to the bacterial suspensions at a ratio of 1:1. The bacteria/lactose suspension is considered the fresh product in the following. Three ml of the fresh product was placed into sterile glass vials and then freeze-dried with the Advantage EL (Virtis, USA) freeze-dryer. The freeze-drying protocol was as follows: freezing rate of 0.04–0.12°C min^{-1} to –40°C and drying temperature of –20°C at 0.15 mbar for 18 h. After the process, the vials were sealed under 0.15 mbar. For the determination of the viable cells and for the bioassays, the freeze-dried product was made up to the original volume by adding sterile de-ionised water.

2.3. Determination of the number of viable cells

Viable cells of fresh and freeze-dried cells of Pfl53 were enumerated in liquid culture using serial dilutions. 120 µl of TSB was pipetted under sterile conditions into 96-well microtitre plates (Rotilabo, Roth). 30 µl of the Pfl53 suspension was added and 1:5 dilution steps were performed. Afterwards, the plates were incubated for at least 48 h at 25°C. Four replicates were carried out per treatment. The number of turbid wells was counted and the Most Probable Number (MPN) procedure was used to enumerate the viable cells (Most Probable Number Calculator Version 4.04, 1996, Albert J. Klee – Risk Reduction Engineering Laboratory, United States Environmental Protection Agency, Cincinnati, Ohio, USA). The survival rate was calculated by dividing the MPN after freeze-drying by the MPN before freeze-drying multiplied by 100.

2.4. Influence of fermentation time

The experiment was performed in the 4-l fermenter. To follow the growth curve of the bacteria and determine the transition from exponential to stationary phase, samples were taken every four hours. The cell density (OD) was measured at 595-nm using a SPECTRA Mini AP (Tecan, Switzerland). A 20% (w/v) solution of lactose was added in ratio of 1:1 to the cell suspension and then the samples were freeze-dried as described previously. MPN was determined before and after freeze-drying. The experiment was time independent repeated five times.

2.5. Influence of the growth media

Three media were tested: KB, a modified TSB (TSB½) in which casein peptone pancreatic (Merck) and soy peptone (Merck) were reduced to half of the original concentration, and a mineral medium (DF) containing an organic source of carbon (1.5 g citric acid monohydrate (Roth), 1.5 g D(+)-glucose monohydrate (Merck), 10 µg H$_3$BO$_3$ (Roth), 7.5 g Na$_2$HPO$_3$ × 2 H$_2$O (Merck), 1 mg FeSO$_4$ × 7 H$_2$O (Merck), 2 g (NH$_4$)$_2$SO$_4$ (Merck), 4 g KH$_2$PO$_4$ (Roth) and 0.2 g MgSO$_4$ × 7 H$_2$O (Merck) in 1000 ml de-ionised water). After inoculation, the flasks were incubated for 16 h at 28°C. MPN were determined before and after freeze-drying. The experiment was time independent repeated four times.

2.6. Influence of growth temperature
Four incubation temperatures 20, 25, 30, and 37°C were compared. After inoculation, the flasks were shaken for 16 h before harvest. MPN was determined before and after freeze-drying. The experiment was time independent repeated three times.

2.7. Influence of mild heat shock
The heat treatment was conducted on late logarithmic growing cells. After inoculation, the flasks were shaken for 15 h at 28°C, followed by incubation at temperatures of 35, 40 or 45°C for one hour. One sample was left at 28°C as control. MPN was determined before and after freeze-drying. The experiment was time independent repeated three times.

2.8. Influence of pH changes
The experiment was performed in the fermenter. After 15 h fermentation, the pH was changed by addition of 1 M HCl (pH reduced successively from 6 to 5 and 4) or 1 M NaOH (pH increased successively from 8 to 9 and 10). After 30 min at the indicated pH, the sample was taken and the pH was changed to the next unit. MPN was determined before and after the freeze-drying process. The experiment was time independent repeated three times.

2.9. Bioassay
The pathogen *B. cinerea* was grown on modified Czapek Dox agar (30 g skim milk powder (Saliter, Germany), 3 g $NaNO_3$, 1 g K_2HPO_4 × 3 H_2O (Merck), 0.5 g $MgSO_4$ × 7 H_2O, 0.5 g KCl, 0.01 g Iron (II)-sulphate × 7 H_2O (Merck), and 18 g agar-agar brought to 1000 ml with de-ionised water) Petri dishes (diameter 9 cm) for 3 weeks at 20°C. Conidial suspensions were prepared by flooding the plates with 0.1% malt extract solution (Merck), gently scraping the mycelium with a spatula and filtering through three layers of muslin in a glass funnel. The concentration of the resulting conidial suspension was counted in a Thoma haemocytometer and adjusted to a concentration of 5×10^5 conidia ml^{-1}. Plants of *V. faba* cv'con Amore' were grown in commercial potting substrate. Compound leaves with four leaflets each of ca. 6-week-old plants were collected. Three compound leaves were placed together on a steel wire mesh with soaked filter paper underneath, in 20 × 20 × 5 cm plexi glass boxes covered with a translucent lid (Gerda GmbH, Schwem, Germany). One ml of Pf153 suspension was sprayed on each leaf surface using an air-brush sprayer. Inoculation with the pathogen followed immediately by spraying 1 ml conidial suspension of *B. cinerea* on the upper side of the leaf. The boxes were then placed in an incubator at 20°C with a day/night cycle of 16/8 h for five days. Both fresh and freeze-dried cells of Pf153 were tested. In the comparison between fresh and freeze-dried cells, the number of viable freeze-dried cells was equalised to the fresh cell value (Table 1). In the bioassay with freeze-dried cells, the viable cells were 5×10^8 MPN ml^{-1}. Water and a 0.2% (w/v) water suspension of Euparen (active ingredient: 50% Tolylfluanid, WG) were used as control treatments. For the

bioassays the disease severity was rated by estimating the affected percent leaf area using the following rating scale: 0 = No lesions, 1 = 0 to 1%, 2 = 2 to 5%, 3 = 6 to 10%, 4 = 11 to 25%, 5 = 26 to 50%, and 6 = 51 to 100%. The affected percent leaf area (APLA) was calculated using Equation (1) (Bora et al., 2004).

$$APLA = \frac{\Sigma(rating\ number \times number\ of\ leaves\ in\ the\ rating)}{total\ number\ of\ leaves \times highest\ rating} \tag{1}$$

Four time independent repetitions for the media assay and three independent repetitions for the temperature assay were set up with four boxes per treatment.

2.10. Statistical analysis

Data were statistically analysed with the software SAS System for Windows v9.3. Experiments on freeze-drying of Pf153 were analysed using the generalised linear model. The Shapiro-Wilk test was applied for the normality test. The homogeneity of variance was proven by the Levene test ($P < 0.1$). For separation of the means, log10 transformed data were compared with the Student-Newman-Keuls test (SNK) ($P < 0.05$). Because no homogeneity of variance or normality was achieved in all of

Table 1. Viability (MPN ml^{-1}), before and after freeze-drying, of *P. fluorescens* Pf153 grown in different conditions.

Growth media	Fresh cells	Freeze-dried cells	Survival rate (%)
TSB½	6.3 ± 3.6a	3.3 ± 1.6b	63 ± 39
DF	6.5 ± 4.0a	4.0 ± 3.0a	63 ± 26
KB	4.7 ± 2.1a	3.4 ± 1.3a	84 ± 48
Growth temperature			
20	19.5 ± 8.9a	7.8 ± 1.1b	49 ± 27
25	7.8 ± 2.0a	7.7 ± 2.8a	106 ± 58
30	6.4 ± 3.9a	5.4 ± 1.2a	115 ± 75
37	8.1 ± 5.2a	2.3 ± 0.6b	34 ± 14
Mild heat shock [°C]			
28	6.7 ± 2.9a	3.0 ± 1.5b	45 ± 15
35	3.8 ± 1.4a	3.9 ± 2.3a	124 ± 87
40	4.9 ± 1.8a	3.3 ± 1.1a	66 ± 15
45	4.4 ± 4.3a	2.7 ± 0.7a	83 ± 58
pH			
4	6.3[§]± 5.8a	2.4[§] 1.9a	39 ± 26
5	4.0 ± 1.3a	2.4 ± 1.3b	66 ± 38
6	4.6 ± 3.0a	3.3 ± 1.5a	90 ± 67
7	4.4 ± 1.7a	6.3 ± 3.9a	145 ± 80
8	3.9 ± 1.4a	6.3 ± 3.1a	159 ± 38
9	7.7 ± 3.3a	7.1 ± 2.5a	110 ± 77
10	2.1[§]± 2.4a	0.7[§]± 0.4a	31 ± 11

Note: All MPN values ×10^8 if not otherwise described. [§]×10^4; [$]×10^6. Initial suspension optical density at 595-nm of 0.95 ± 0.02; the data are means (±SD) of three (four for media) time independent repetitions. Means of the data in one line followed by the same letter are not significantly different following Student-Newman Keuls test ($P < 0.05$).
Mild heat shock: heat treatment applied for one hour at late logarithmic phase.
pH: acid/base treatment was applied after 15 h growing by adding acid or base to the next unit and taking the sample after 30 min.

the bioassays, the non-parametric test of Kruskal-Wallis was chosen. By the exact Methods in the NPAR1WAY procedure (two-sided), the samples were compared pair wise (Wilcoxon, exact $P < 0.05$).

3. Results

3.1. Influence of fermentation time

The OD value was monitored to follow the biomass increase over time; it showed that after 16 h the cells exceeded the log phase to enter the stationary phase (Figure 1). The count of viable cells showed an increase from 1.3×10^8 MPN ml^{-1} (inoculation) to 7.9×10^9 MPN ml^{-1} (after 16 h) and a further increase at 24 h to 2.5×10^{10} MPN ml^{-1} (data not shown). The viability after freeze-drying increased with fermentation time. At 16 and 20 h the survival rates were 115% and 117% respectively, and decreased at 24 h to 37%. Therefore the bacteria were harvested at 16 h in all following experiments; this harvesting time yielded a good mass quantity and a high viability after conservation process.

3.2. Influence of the growth media

In preliminary experiments, Pf153 was grown in nine different media, based on three standard media varying in carbon (C) and nitrogen (N) concentrations, and the biomass was monitored (data not shown). Out of them the best C and N concentration of each standard medium was chosen to study their influence on the viability after freeze-drying. From these media, only TSB½ showed a statistical

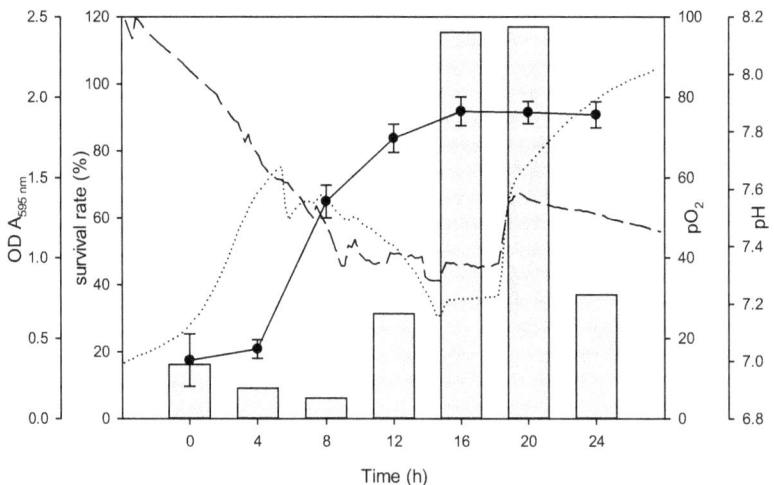

Figure 1. Fermentation data of *P. fluorescens* Pf153 monitored for 24 h at 28°C, 900 rpm and 3.70 Nl min^{-1} air (data from repetition 4; pH as dotted line and pO$_2$ as long dash line), biomass production (OD, solid line with dots) and survival rates (columns) calculated after freeze-drying.

difference (Table 1) between the viability of fresh and freeze-dried product. The survival rate was 63% for TSB½, 63% for DF and 84% for KB.

3.3. Influence of growth temperature

After 16 h fermentation the OD values from the four temperatures were included between 0.881 at 20°C and 1.228 for the cells grown at 37°C (ANOVA, F = 9.17; df = 3,8; P = 0.0057; statistically different by SNK for 37°C). The MPN ml^{-1} values were also the lowest at 20°C (4.4 × 10^9 MPN ml^{-1}), and were maximum at 30°C (1.4 × 10^{10} MPN ml^{-1}) followed by 37°C (1.1 × 10^{10} MPN ml^{-1}) and 25°C (8.2 × 10^9 MPN ml^{-1}) but not statistically different (ANOVA, F = 1.47; df = 3,8; P = 0.2947). As shown in Table 1, the fermentation temperature influences the viability of the cells after freeze-drying. In fact the cells grown at 20°C and 37°C showed a minor survival rate compared with the cells grown at 25 and 30°C. A higher growing temperature reduced the number of viable cells to less than 35%.

3.4. Influence of mild heat shock

In this experiment a temperature increase at the end of the log phase influenced the survival of Pf153 after freeze-drying. Results in Table 1 show that a mild heat shock of one hour significantly influenced the viability during the freeze-drying process. The survival rate increased to 124% at 35°C, but was reduced to at least 40% at higher temperature treatments.

3.5. Influence of pH changes

The pH monitoring data are shown in Figure 1. The pH ranged between 7.06 ± 0.15 (fermentation beginning) and 8.04 ± 0.12 (fermentation end). To study the influence of a pH change on survival rate, after 15 h fermentation (pH value 7.24 ± 0.10) HCl or NaOH was added to the fermenter. pH values between 6 and 9 did not significantly influence the survival rate during freeze-drying of Pf153, the values ranged from 61% for pH 5 to more than 100% for pH 8 (Table 1). A pH of 10 or 4 reduced the number of fresh living cells and resulted in additional instability during freeze-drying (survival of 31 and 39% respectively).

3.6. Bioassay

Cell suspensions with similar bacterial concentrations (MPN ml^{-1}) and grown in different media, were tested against *B. cinerea* as both fresh and freeze-dried products. The cells produced in the three media were all able to reduce the disease severity significantly with a reduction of at least 45% (KB) compared to the water control, with no differences in efficacy between media (Table 2). The freeze-dried cells showed the same ability to reduce the disease as the fresh cells, resulting in a reduction of at least 43% (TSB½) against water control. The differences in fermentation media composition did not significantly influence the activity of Pf153, either fresh or freeze-dried. However the efficacy was influenced by the growing temperature (Table 3). According to Table 3 the efficacy of Pf153 grown at 20°C was significantly higher compared to the other growth temperatures (25, 30 and 37°C), increasing to 52.9% compared with 31.7% of the next best value (25°C). As shown in Table 3, a mild heat shock did not improve the efficacy of the bacterial antagonist; no significant differences were observed.

Table 2. Influence of treatments of fresh and freeze-dried *P. fluorescens* Pf153 cultivated in different growing media on severity (%) of *B. cinerea* infections on detached leaves of *V. faba.*

Treatment	Fresh cells	Freeze-dried cells
Water control	75.7 ± 14.5a	73.6 ± 16.7a
TSB½	40.3 ± 3.4 b	41.7 ± 10.0b
DF	40.6 ± 1.3 b	40.3 ± 8.6 b
KB	41.3 ± 3.8 b	40.6 ± 9.4 b
Euparen	19.8 ± 4.0 c	19.4 ± 4.4 c

Note: MPN ml^{-1} before and after freeze-drying are equal (TSB½: 6 × 10^8; DF: 6 × 10^8; KB: 5 × 10^8 MPN/ml).
Means of the data in a column followed by the same letter are not significantly different following Wilcoxon test ($P < 0.05$).
The data are means (±SD) of four independent experiments.
Severity was assessed five days after inoculation.

4. Discussion

The key objectives for the development of BCAs are high viability, storability and tolerance to variable environmental conditions. The quality of formulated micro-organisms can be improved by the physiological quality (Teixidó et al., 2005). Fermentation medium, growing temperature, time and pH and aeration rate play as well an important role (Ashofteh, Ahmadzadeh, & Fallahzadeh-Mamaghani, 2009; Hynes & Boyetchko, 2006; Spadaro & Gullino, 2005). Therefore, in the present study, we investigated the effect of diverse growth conditions on the survival and the efficacy of *P. fluorescens* Pf153 after freeze-drying.

During liquid fermentation of Pseudomonads nutrients are consumed while metabolites are produced and cells enter the stationary phase preparing for survival

Table 3. Influence of the growth temperature and mild heat shock treatment on the efficacy of freeze-dried *P. fluorescens* Pf153 cells against *B. cinerea* on detached leaves of *V. faba.*

Fermentation temperature	Treatment	Mean severity (%)
	Water	75.9 ± 7.6 a
	20°C	35.6 ± 5.8 c
	25°C	52.3 ± 14.0b
	30°C	55.6 ± 5.0 b
	37°C	56.0 ± 12.9b
	Euparen	14.4 ± 1.6 d
Mild heat shock	Water	77.8 ± 12.0a
	28°C	38.0 ± 8.5 b
	35°C	41.7 ± 7.3 b
	40°C	38.9 ± 7.7 b
	45°C	40.7 ± 2.9 b
	Euparen	18.1 ± 2.8 c

Note: all suspensions contain 5 × 10^8 MPN ml^{-1}. Means of the data for each treatment followed by the same letter are not significantly different following Wilcoxon test ($P < 0.05$). The data are means (±SD) of three independent experiments.
Mild heat shock: heat treatment applied for one hour at late logarithmic phase.
Severity was assessed five days after inoculation.

(Slininger, VanCauwenberge, Shea-Wilbur, & Bothast, 1997); cell stability can be optimised through fermentation conditions and harvesting time (Slininger et al., 1996). Heckly (1985) postulated, that in the late log or stationary phase cultures are less sensitive to the freeze-drying process than young cells (Heckly, 1985). In our experiments in the fermenter, cells of Pf153 harvested during the initial growing phase had a significantly lower viability than those harvested at the end of the log or in the stationary phase. Pf153 harvested after 24 h (about 8 h in the stationary phase) had a lower survival rate than cells harvested at 16 h (end of the log phase). Slininger et al. (1996, 1997) using encapsulated *P. fluorescens* 2-79 got similar results. In their experiments younger cells (24–48 h) survived drying better than older cells (72–96 h) which could have been caused by different cell protection mechanisms during culturing (Slininger et al., 1996, 1997). In contrast, the survival rate after freeze-drying of *P. putida* KT2440 was higher for cells in the stationary phase than for cells in the log phase underlying that optimal harvesting time is largely organism dependent (Muñoz-Rojas et al., 2006).

Pf153 grown at different temperatures showed differences in the survival rates after freeze-drying. The viability of cells grown at 25 or 30°C was not reduced after freeze-drying but when cultivated at 20 or 37°C the number of viable cells was significantly lower. The impact of fermentation temperature on survival after freezing was also shown for *P. putida* GR 12-2 by Sun, Griffith, Pasternak, and Glick (1995). It is generally accepted that lower growing temperatures favour fatty acid production (Hansson & Dostálek, 1988), which may account for the difference in survival rates. Fatty acid composition and oxidation degree appear to be related to the survival of cells during the freeze-drying process (Carvalho et al., 2003).

In addition to cultivation temperature, fermentation media can also influence the survival after freeze-drying through accumulated compatible solutes and by altering the cell membrane (Carvalho et al., 2004). In our studies with Pf153 highest survival rate was achieved by cells grown in KB. Because in all tested media the stationary phase was reached after 16 hours (data not shown) fermentation time it can be excluded that the different survival rates were caused by different cell age. The KB medium contains glycerol which is reported to improve cell viability during freeze-drying (Heckly, 1985). However, further experiments need to be conducted to identify the medium component responsible for higher survival rates after freeze-drying. Sugars may possibly play an important role. Sartori, Nesci, and Etcheverry (2012) proved that an addition of 10% sucrose to the growing media increased viability of *Bacillus amyloliquefaciens* and *Microbacterium oleovorans* significantly during freeze-drying; a molasses and soy powder medium gave the best biomass production, a high viability and an improved performance for both bacteria (Sartori et al., 2012).

Bacteria have developed adaptive strategies to counter the challenges of changing environments and to survive under stress conditions (Abee & Wouters, 1999). There is a general consensus that disadvantageous conditions during microbial growth can induce tolerance responses (Morgan et al., 2006). Different abiotic stresses like thermal (heat or cold) and non-thermal (i.e. pH, salinity) stressors, can affect the cell structure (Panoff, Thammavongs, Gueguen, & Boutibonnes, 1998) and may induce the production of stress proteins (Ashofteh et al., 2009; Khare & Arora, 2011). Sub-lethal stresses can induce tolerance to a more extreme stress and may cross-protect against other types of stress (Teixidó et al.,

2005). In our experiments raising the temperature by 7°C for ca. one hour increased the survival after the freeze-drying process by almost 80%. The viability was still improved at higher temperatures (40 and 45°C) but less than at 35°C. An increased resistance to freeze-drying due to heat shock is also reported for other BCAs. Palmfeldt et al. (2003) could double the survival of *P. chlororaphis* MA 100, when cells were exposed to a heat treatment for 15 min at 34.5°C (Palmfeldt et al., 2003). The cultivability of freeze-dried *Lactococcus lactis* ssp. *lactis* cells was significantly improved by a heat shock of 25 min at 42°C. For cell protection, heat shock protein expression was of main significance, but the membrane lipid composition also changed in response to the temperature stress (Broadbent & Lin, 1999). The time in which the heat stress is applied can also be important. In our studies heat stress was applied at the end of the log phase. Carvalho et al. (2004) reported that the application of a heat shock increased the survival of log phase lactic acidic bacteria (LAB) during spray-drying (Carvalho et al., 2004). Saarela et al. (2004) reported that cells of *Bifidobacterium animalis*, stressed during the stationary phase are more resistant to various stresses than cells in the log phase (Saarela et al., 2004).

Acid adaptation or shock during fermentation may produce cells with a different physiological state and therefore with tolerances to other stresses. Acid-adapted cells are cells exposed to a gradual decrease in environmental pH, whereas acid-shocked cells are those exposed to an abrupt pH shift from high to low pH (Abee & Wounters, 1999). In our experiments, pH change reduced survival rates to less than 40% during freeze-drying only at extreme pH values of 4 and 10. Cells adapted to low pH seemed to have different sensitivities to further stress. When *Lactobacillus reuteri* was grown at a suboptimal pH of 5, the cell viability after freeze-drying increased when harvested after 2.5 h in the stationary phase (Palmfeld & Hanh-Hägerdal, 2000). *Lb. bulgaricus* ND02 adapted to a lower pH (reduction of 0.6 points) improved the survival of the cells after lyophilising of about 17% (Shao, Gao, Guo, & Zhang, 2014). In studies with LAB, a stress caused by a reduced pH could enhance viability after freeze-drying as a result of adaptation (Capela, Hay, & Shah, 2006).The above mentioned results indicate that pre-treatment of cells before freeze-drying protected them from the stress generated during the drying process (Shao et al., 2014).

Biotic and abiotic factors can influence the activity of bacterial antagonists (Borowicz & Omer, 2000; Dickie & Bell, 1995; Slininger et al., 1996). Dickie and Bell (1995) pointed out that the growth history of three antagonistic strains against *Agrobacterium vitis* significantly influenced the activity in laboratory tests. The medium composition, growing temperature and pH of the media had different influences on the antagonistic capacity of the tested BCA (Dickie & Bell, 1995).

In our bioassay, in which cells of Pf153 were tested against *B. cinerea* on *V. faba* leaves, no differences in the efficacy between the cells grown in the three tested media were observed. The efficacy of fresh fermented cells ranged between 45.4 and 46.8%. Comparable results were also reported by Fuchs et al. (2000). In artificial soils, strain Pf153 was able to reduce the disease index of black root rot on cucumber from 81% to less than 18% regardless the bacterial growth medium. However, strain Pf153 showed differences in biocontrol efficacy when the tests were performed in natural soils (Fuchs et al., 2000). Borowicz and Saad Omer (2000) reported that the growth-promoting activity of *Pseudomonas* spp. on cucumber could be improved or reduced by culturing the bacteria in different media. For example *P. fluorescens* PPs21 positively affected plant growth when cultivated in liquid media, while a change in media status (to solid) produced an inhibition of cucumber growth (Borowicz & Omer, 2000). The *in vitro* and *ad planta* efficacy of *P. fluorescens* strain UTPF61 against *Sclerotinia sclerotiorum* was influenced by different mineral contents and their combinations in the fermentation media (Ashofteh et al., 2009). Peighami-Ashnaei, Sharifi-Tehrani, Ahmadzadeh, and Behboudi et al. (2009) reported that the biocontrol efficacy of *P. fluorescens* P-35 against *B. cinerea* on apples, grown in a medium containing molasses and yeast extract, was significantly better than that of other media tested in the experiment (Peighami-Ashnaei et al., 2009). The media we tested differed in C and N sources, C:N ratios and starting pH value, but they did not influence the efficacy in the bioassay. Therefore, it is suggested that the efficacy of Pf153 is less media dependent.

Additionally, when freshly harvested and freeze-dried cells were tested against *B. cinerea* on *V. faba*, no negative impact of freeze-drying on the efficacy of Pf153 was observed. However, Khare and Arora (2011) reported that freeze-dried fluorescent *Pseudomonas* EKi cells exhibited a reduced plant growth promotory activity of chickpea in *Macrophomina phaseolina* infested soil compared to non-treated cells.

To enhance root colonisation and efficacy of biocontrol fluorescent pseudomonas, Gu and Mazzola (2001) proposed exposing the bacteria *in vitro* to conditions that induce resistance to stress conditions encountered in natural soil environments (Gu & Mazzola, 2001). In 1991 Deacon suggested that the following effort to develop a BCA failed because the antagonist selections were done *in vitro* but the potential antagonist was ecologically unsuited to the environment where the pathogens grow (Deacon, 1991). Therefore, in our experiments we compared different fermentation temperatures for the production of Pf153. About 20°C was included because at this temperature the fungal disease is active. The results indicate that the efficacy was influenced significantly by the growing temperature. The best results were obtained when the fermentation and bioassay temperature were similar. Possibly, the bacteria take less time to adapt to the new environment and can therefore be effective in a shorter period of time. Low growing temperatures positively influenced the efficacy of Pf153. A positive effect of low growing temperatures is also described by Dickie and Bell (1995), where the inhibition of the bacterial antagonist against *Agrobacterium vitis* was stronger at 15°C than 30°C (Dickie & Bell, 1995).

5. Conclusion

The results demonstrate that by optimising media composition, growth conditions and fermentation time, the survival of *P. fluorescens* Pf153 during the freeze-drying process can be enhanced. For Pf153 freeze-drying was shown to be an efficient conservation method resulting in high viability and an efficacy comparable to fresh cells. The resulting product is easy to handle and from previous experiments has shown good storability. Furthermore, the efficacy of freeze-dried cells of Pf153 against *B. cinerea* can be increased by optimising growth temperature.

Disclosure statement

No potential conflict of interest was reported by the authors.

Funding

This research was financially supported by Safecrop Centre, funded by Fondo per la Ricerca, Autonomous Province of Trento. Monika Maurhofer (ETH, Switzerland) provided the *P. fluorescens* strain Pf153.

References

Abadias, M., Benabarre, A., Teixidó, N., Usall, J., & Viñas, I. (2001). Effect of freeze drying and protectants on viability of the biocontrol yeast *Candida sake*. *International Journal of Food Microbiology, 65*, 173–182. doi:10.1016/s0168-1605(00)00513-4

Abee, T., & Wouters, J. A. (1999). Microbial stress response in minimal processing. *International Journal of Food Microbiology, 50*, 65–91. doi:10.1016/s0168-1605(99)00078-1

Ashofteh, F., Ahmadzadeh, M., & Fallahzadeh-Mamaghani, V. (2009). Effect of mineral components of the medium used to grow biocontrol strain UTPF61 of *Pseudomonas fluorescens* on its antagonistic activity against *Sclerotinia* wilt of sunflower and its survival during and after formulation process. *Journal of Plant Pathology, 91*, 607–613.

Berny, J. F., & Hennebert, G. L. (1991). Viability and stability of yeast-cells and filamentous fungus spores during freeze-drying – Effects of protectants and cooling rates. *Mycologia, 83*, 805–815. doi:10.2307/3760439

Bora, T., Ozaktan, H., Gore, E., & Aslan, E. (2004). Biological control of *Fusarium oxysporum* f. sp melonis by wettable powder formulations of the two strains of *Pseudomonas putida*. *Journal of Phytopathology, 152*, 471–475. doi:10.1111/j.1439-0434.2004.00877.x

Borowicz, J. J., & Omer, Z. S. (2000). Influence of rhizobacterial culture media on plant growth and on inhibition of fungal pathogens. *Biocontrol, 45*, 355–371. doi:10.1023/a:1009954802552

Broadbent, J. R., & Lin, C. (1999). Effect of heat shock or cold shock treatment on the resistance of *Lactococcus lactis* to freezing and lyophilization. *Cryobiology, 39*, 88–102. doi:10.1006/cryo.1999.2190

Capela, P., Hay, T. K. C., & Shah, N. P. (2006). Effect of cryoprotectants, prebiotics and microencapsulation on survival of probiotic organisms in yoghurt and freeze-dried yoghurt. *Food Research International, 39*, 203–211. doi:10.1016/j.foodres.2005.07.007

Carvalho, A. S., Silva, J., Ho P., Teixeira, P., Malcata, F. X., & Gibbs, P. (2003). Effect of various growth media upon survival during storage of freeze-dried *Enterococcus faecalis* and *Enterococcus durans*. *Journal of Applied Microbiology, 94*, 947–952. doi:10.1046/j.1365-2672.2003.01853.x

Carvalho, A. S., Silva, J., Ho, P., Teixeira, P., Malcata, F. X., & Gibbs, P. (2004). Relevant factors for the preparation of freeze-dried lactic acid bacteria. *International Dairy Journal, 14*, 835–847. doi:10.1016/j.idairyj.2004.02.001

Costa, E., Usall, J., Teixido, N., Torres, R., & Vinas, I. (2002). Effect of package and storage conditions on viability and efficacy of the freeze-dried biocontrol agent *Pantoea agglomerans* strain CPA-2. *Journal of Applied Microbiology, 92*, 873–878. doi:10.1046/j.1365-2672.2002.01596.x

Deacon, J. W. (1991). Significance of ecology in the development of biocontrol agents against soil-borne plant pathogens (Abstract). *Biocontrol Science and Technology, 1*, 5–20. doi:10.1080/09583159109355181

Dickie, G. A., & Bell, C. R. (1995). A full factorial analysis of nine factors influencing *in vitro* antagonistic screens for potential biocontrol agents. *Canadian Journal of Microbiology, 41*, 284–293. doi:10.1139/m95-039

Fuchs, J. G., & Defago, G. (1991). Protection of cucumber plants against black root rot caused by *Phomopsis sclerotioides* with rhizobacteria. *Bulletin SROP, 14*, 57–62.

Fuchs, J. G., Moënne-Loccoz, Y., & Défago, G. (2000). The laboratory medium used to grow biocontrol *Pseudomonas* sp Pf153 influences its subsequent ability to protect cucumber from black root rot. *Soil Biology & Biochemistry, 32,* 421–424. doi:10.1016/s0038-0717(99)00169-8

Gu, Y. H., & Mazzola, M. (2001). Impact of carbon starvation on stress resistance, survival in soil habitats and biocontrol ability of *Pseudomonas putida* strain 2C8. *Soil Biology & Biochemistry, 33,* 1155–1162. doi:10.1016/s0038-0717(01)00019-0

Hansson, L., & Dostálek, M. (1988). Effect of culture conditions on mycelial growth and production of gamma-linolenic acid by the fungus *Mortierella ramanniana. Applied Microbiology and Biotechnology, 28,* 240–246.

Heckly, R. J. (1985). Principles of preserving bacteria by freeze-drying. *Developments in Industrial Microbiology, 26,* 379–396.

Hynes, R. K., & Boyetchko, S. M. (2006). Research initiatives in the art and science of biopesticide formulations. *Soil Biology & Biochemistry, 38,* 845–849. doi:10.1016/j.soilbio.2005.07.003

Khare, E., & Arora, N. K. (2011). Physiologically stressed cells of Fluorescent Pseudomonas EKi as better option for bioformulation development for management of charcoal rot caused by *Macrophomina phaseolina* in field conditions. *Current Microbiology, 62,* 1789–1793. doi:10.1007/s00284-011-9929-x

King, E. O., Ward, M. K., & Raney, D. E. (1954). Two simple media for the demonstration of pyocyanin and fluorescin. *Journal of Laboratory and Clinical Medicine, 44,* 301–307.

Liu, Z., Wei, H., Li, Y., Li, S., Luo, Y., Zhang, D., & Ni, L. (2014). Optimization of the spray drying of a *Paenibacillus polymyxa*-based biopesticide on pilot plant and production scales. *Biocontrol Science and Technology, 24,* 426–435. doi:10.1080/09583157.2013.868865

Manikandan, R., Saravanakumar, D., Rajendran, L., Raguchander, T., & Samiyappan, R. (2010). Standardization of liquid formulation of *Pseudomonas fluorescens* Pf1 for its efficacy against *Fusarium* wilt of tomato. *Biological Control, 54,* 83–89. doi:10.1016/j.biocontrol.2010.04.004

Mathivanan, N., Prabavathy, V. R., & Vijayanandraj, V. R. (2005). Application of talc formulations of *Pseudomonas fluorescens* migula and *Trichoderma viride* pers. Ex SF Gray decrease the sheath blight disease and enhance the plant growth and yield in rice. *Journal of Phytopathology, 153,* 697–701. doi:10.1111/j.1439-0434.2005.01042.x

Meena, B. (2010). Survival and effect of *Pseudomonas fluorescens* formulation developed with various carrier materials in the management of late leaf spot of groundnut. *International Journal of Plant Protection, 3,* 200–202.

Miyamoto-Shinohara, Y., Imaizumi, T., Sukenobe, J., Murakami, Y., Kawamura, S., & Komatsu, Y. (2000). Survival rate of microbes after freeze-drying and long-term storage. *Cryobiology, 41,* 251–255. doi:10.1006/cryo.2000.2282

Montesinos, E. (2003). Development, registration and commercialization of microbial pesticides for plant protection. *International Microbiology, 6,* 245–252. doi:10.1007/s10123-003-0144-x

Morgan, C. A., Herman, N., White, P. A., & Vesey, G. (2006). Preservation of micro-organisms by drying; A review. *Journal of Microbiological Methods, 66,* 183–193. doi:10.1016/j.mimet.2006.02.017

Muñoz-Rojas, J., Bernal, P., Duque, E., Godoy, P., Segura, A., & Ramos, J. L. (2006). Involvement of cyclopropane fatty acids in the response of *Pseudomonas putida* KT2440 to freeze-drying. *Applied and Environmental Microbiology, 72,* 472–477. doi:10.1128/aem.72.1.472-477.2006

Nandakumar, R., Babu, S., Viswanathan, R., Raguchander, T., & Samiyappan, R. (2001). Induction of systemic resistance in rice against sheath blight disease by *Pseudomonas fluorescens. Soil Biology & Biochemistry, 33,* 603–612. doi:10.1016/s0038-0717(00)00202-9

Paau, A. S. (1988). Formulations useful in applying beneficial microorganisms to seeds. *Trends in Biotechnology, 6,* 276–279. doi:10.1016/0167-7799(88)90124-2

Palmfeldt, J., & Hahn-Hägerdal, B. (2000). Influence of culture pH on survival of *Lactobacillus reuteri* subjected to freeze-drying. *International Journal of Food Microbiology, 55,* 235–238. doi:10.1016/s0168-1605(00)00176-8

Palmfeldt, J., Rådström, P., & Hahn-Hägerdal, B. (2003). Optimisation of initial cell concentration enhances freeze-drying tolerance of *Pseudomonas chlororaphis. Cryobiology, 47,* 21–29. doi:10.1016/s0011-2240(03)00065-8

Panoff, J. M., Thammavongs, B., Gueguen, M., & Boutibonnes, P. (1998). Cold stress responses in mesophilic bacteria. *Cryobiology, 36,* 75–83. doi:10.1006/cryo.1997.2069

Peighami-Ashnaei, S., Sharifi-Tehrani, A., Ahmadzadeh, M., & Behboudi, K. (2009). Interaction of different media on production and biocontrol efficacy of *Pseudomonas fluorescens* P-35 and *Bacillus subtilis* B-3 against grey mould of apple. *Journal of Plant Pathology, 91,* 65–70.

Rabindran, R., & Vidhyasekaran, P. (1996). Development of a formulation of *Pseudomonas fluorescens* PfALR2 for management of rice sheath blight. *Crop Protection, 15,* 715–721. doi:10.1016/s0261-2194(96)00045-2

Radja Commare, R., Nandakumar, R., Kandan, A., Suresh, S., Bharathi, M., Raguchander, T., & Samiyappan, R. (2002). Pseudomonas fluorescens based bio-formulation for the management of sheath blight disease and leaffolder insect in rice. *Crop Protection, 21,* 671–677. doi:10.1016/S0261-2194(02)00020-0

Saarela, M., Rantala, M., Hallamaa, K., Nohynek, L., Virkajarvi, I., & Mättö, J. (2004). Stationary-phase acid and heat treatments for improvement of the viability of probiotic lactobacilli and bifidobacteria. *Journal of Applied Microbiology, 96,* 1205–1214. doi:10.1111/j.1365-2672.2004.02286.x

Sartori, M., Nesci, A., & Etcheverry, M. (2012). Production of *Fusarium verticillioides* biocontrol agents, *Bacillus amyloliquefaciens* and *Microbacterium oleovorans,* using different growth media: Evaluation of biomass and viability after freeze-drying. *Food Additives and Contaminants Part A Chemistry Analysis Control Exposure & Risk Assessment, 29,* 287–292. doi:10.1080/19440049.2011.563369

Selvaraj, S., Ganeshamoorthi, P., Anand, T., Raguchander, T., Seenivasan, N., & Samiyappan, R. (2014). Evaluation of a liquid formulation of *Pseudomonas fluorescens* against *Fusarium oxysporum* f. sp *cubense* and *Helicotylenchus* multicinctus in banana plantation. *Biocontrol, 59,* 345–355. doi:10.1007/s10526-014-9569-8

Shao, Y., Gao, S., Guo, H., & Zhang, H. (2014). Influence of culture conditions and preconditioning on survival of *Lactobacillus delbrueckii* subspecies *bulgaricus* ND02 during lyophilization. *Journal of Dairy Science, 97,* 1270–1280. doi:10.3168/jds.2013-7536

Slininger, P. J., Cauwenberge, J.E.V., Bothast, R. J., Weller, D. M., Thomashow, L. S., & Cook, R. J. (1996). Effect of growth culture physiological state, metabolites, and formulation on the viability, phytotoxicity, and efficacy of the take-all biocontrol agent Pseudomonas fluorescens 2-79 stored encapsulated on wheat seeds. *Applied Microbiology and Biotechnology, 45,* 391–398. doi:10.1007/s002530050701

Slininger, P. J., VanCauwenberge, J. E., Shea-Wilbur, M. A., & Bothast, R. J. (1997). Impact of liquid culture physiology, environment, and metabolites of biocontrol agent qualities. In G. J. Boland & L. D. Kuykendall (Eds.), *Plant-microbe interactions and biological control* (pp. 329–353). New York, NY: Marcel Dekker.

Smith, D., & Onions, A. H. S. (1983). *The preservation and maintenance of living fungi.* Kew, UK: Commonwealth Mycological Institute.

Spadaro, D., & Gullino, M. L. (2005). Improving the efficacy of biocontrol agents against soilborne pathogens. *Crop Protection, 24,* 601–613. doi:10.1016/j.cropro.2004.11.003

Stephan D., Bisutti I. L., Matos da Silva A.-P. (2007). Optimisation of the freeze drying process of *Pseudomonas fluorescens* strain CHA0 and Pf 153. *Bulletin SROP, 30,* 511–515.

Sun, X., Griffith, M., Pasternak, J. J., & Glick, B. R. (1995). Low temperature growth, freezing survival, and production of antifreeze protein by the plant growth promoting rhizobacterium Pseudomonas putida GR12-212-2. *Canadian Journal of Microbiology, 41,* 776–784. doi:10.1139/m95-107

Teixidó, N., Cañamás, T. P., Usall, J., Torres, R., Magan, N., & Viñas, I. (2005). Accumulation of the compatible solutes, glycine-betaine and ectoine, in osmotic stress adaptation and heat shock cross-protection in the biocontrol agent *Pantoea agglomerans* CPA-2. *Letters in Applied Microbiology, 41,* 248–252. doi:10.1111/j.1472-765X.2005.01757.x

Thakur, R. P., & Rao, V. P. (2002). Biological control of preharvest kernel infection by *Apergillus flavus* in groundnut, 2001. *(American Phytopathological Society) B&C Tests 17,* P02.

Thakur, R. P., Rao, V. P., & Subramanyam, K. (2003). Influence of biocontrol agents on population density of *Aspergillus flavus* and kernel infection in groundnut. *Indian Phytopathology, 56*, 408–412.

Umesha, S., Dharmesh, S. M., Shetty, S. A., Krishnappa, M., & Shetty, H. S. (1998). Biocontrol of downy mildew disease of pearl millet using *Pseudomonas fluorescens*. *Crop Protection, 17*, 387–392. doi:10.1016/s0261-2194(98)00014-3

Vidhyasekaran, P., Rabindran, R., Muthamilan, M., Nayar, K., Rajappan, K., Subramanian, N., & Vasumathi, K. (1997). Development of a powder formulation of *Pseudomonas fluorescens* for control of rice blast. *Plant Pathology, 46*, 291–297. doi:10.1046/j.1365-3059.1997.d01-27.x

Vidhyasekaran, P., Sethuraman, K., Rajappan, K., & Vasumathi, K. (1997). Powder formulations of *Pseudomonas fluorescens* to control pigeonpea wilt. *Biological Control, 8*, 166–171. doi:10.1006/bcon.1997.0511

Zhao, G., & Zhang, G. (2005). Effect of protective agents, freezing temperature, rehydration media on viability of malolactic bacteria subjected to freeze-drying. *Journal of Applied Microbiology, 99*, 333–338. doi:10.1111/j.1365-2672.2005.02587.x

Influence of fermentation temperature and duration on survival and biocontrol efficacy of *Pseudomonas fluorescens* Pf153 freeze-dried cells (2019).

Bisutti I.L. and Stephan D.

Submitted to Journal of Applied Microbiology

Influence of fermentation temperature and duration on survival and biocontrol efficacy of *Pseudomonas fluorescens* Pf153 freeze-dried cells

I.L. Bisutti[1,2] and D. Stephan[1]

[1] Julius Kühn-Institut, Federal Research Centre for Cultivated Plants, Institute for Biological Control
[2] Humboldt-Universität zu Berlin, Faculty of Life Science, Division Phytomedicine

Abstract

Aim: The aim of this paper was to determine whether the quality of formulated *Pseudomonas fluorescens* Pf153 can be influenced by changes in fermentation conditions. In this study the effect of the duration and temperature of fermentation on shelf life, viability and biocontrol efficacy of freeze-dried cells of *P. fluorescens* Pf153 was investigated.

Methods and Results: Cells of *P. fluorescens* Pf153 were grown at 20°C and 28°C in flasks and fermenter and harvested in the mid-log and the beginning of the stationary phase. The survival during storage of freeze-dried cells was tested at 25°C. Cells fermented at 20°C survived in storage better than those grown at 28°C, irrespective of the harvesting time. Compared to the untreated control, in *in vitro* tests Pf153 was in all production temperature/duration combinations significantly effective against all tested *Botrytis cinerea* strains. But no differences between temperature/duration combinations were found. In bioassay on detached *Vicia faba* leaves, a significant influence of the fermentation temperature/duration combinations on the biocontrol efficacy was found.

Conclusions: These results demonstrate that fermentation parameters have an influence on the performance and quality of a formulated product.

Significance and Impact of the Study: Only limited numbers of biocontrol products based on antagonistic pseudomonads are on the market. This can be attributed to the lack of suitable formulated products with high numbers of viable cells and a good shelf life. Currently, only limited information on the influence of the fermentation on subsequent downstreaming process is available. Within this study, we focused on the influence of the two important parameters fermentation temperature and harvest time on survival, shelf life and biocontrol efficacy of *P. fluorescens* Pf153.

Key words: Agriculture, Production, Pseudomonads, Bioprocessing, Shelf life

Introduction

It requires not only a successful scale-up of the fermentation process but also the effective formulation of the organisms (Lee et al., 2006). In addition, the end product should be easy to handle and store (e.g. at room temperature) and to be easily applied (Angeli et al., 2016). *Pseudomonas* spp. are fast growing bacteria which can colonize plants roots and phyllosphere. They produce a large spectrum of bioactive metabolites like siderophores, volatiles and growth promoting substances, assist the plant in adapting to environmental stresses, are highly competitive with other micro-organism and can suppress soil borne pathogens. *Pseudomonas* spp. also produce antibiotic compounds (Bhattacharjee and Dey, 2014). Several strains of *P. fluorescens* have been shown to antagonize plant pathogens and have thus been considered BCA (Jain and Das, 2016; Manikandan et al., 2010). However, despite the proven potential as biocontrol agents, there are few commercial products based on *Pseudomonas* spp. on the market. Currently there are only two products in the European Union, (https://ec.europa.eu) and nine in the USA (https://www.epa.gov) that list *Pseudomonas* as active ingredients. One of the main reasons is the difficulty to formulate these bacteria for commercial use. *Pseudomonas* spp. are extremely sensitive to environmental stress factors and the available formulations rapidly loose viability (Corrêa et al., 2015; Paulitz and Belanger, 2001).

Several variables affect survival: the fermentation medium, the physiological state at harvest time, the protective materials used during the formulation process and the type of drying technology (Bashan et al., 2014). The formulation process is designed to promote diverse BCA traits like survival in soil, production of antimicrobial compounds and disease suppression effectiveness. This influence can be positive or negative (Fuchs et al., 2000). To retain the biological traits of the formulated final product during extended storage is another major challenge for any formulation (Bashan et al., 2014). Fuchs et al. (2000) also considered that the cultivation conditions prior to formulation can influence the BCA traits positively or negatively.

Among the options for formulation of microorganisms, freeze-drying is considered a gentle dehydration process, suitable to achieve a stable powder formulation. It is the most common technique for drying and storing bacteria and especially suitable for *P. fluorescens* (Mputu, 2014; Stephan et al., 2016). However, due to intrinsic characteristics like cell size and membrane composition the process has to be optimized for each strain. Also, external conditions like presence of nutrients and freezing and drying parameters influence the resistance to lyophilisation and storage. Protective additives help to reduce cells damage during freeze-drying (Tanimomo et al., 2016). Fermentation parameters also influence the cell and therefore its freeze-drying resistance, but also the subsequent biocontrol efficacy. Media composition, for example, influences biomass production (Angeli et al., 2016; He et al., 2008) and the production of secondary metabolites like antibiotics (Gao et al., 2016; He et al., 2008). Shelf-life is one important factor influencing the acceptance of plant protection products. Shelf life of at least one year at room temperature is desired, and the final product should deliver a certain quantity in micro-organism cells or spores even after longer storage (Bashan et al., 2014).

During previous studies, it was found that freeze-dried *P. fluorescens* Pfl53 (Pfl53) cells reduced the infection symptoms of *Botrytis cinerea* on *Vicia faba* leaves. The level of control achieved was influenced by the fermentation temperature (Bisutti et al., 2015). In another study, fermentation media composition influenced the efficacy of Pfl53 against *Phomopsis sclerotioides* mediated black root rot on cucumber (Fuchs et al., 2000). Based on the information that lower fermentation temperature positively influenced the biocontrol activity of Pfl53 against *B. cinerea*, the aim of the present study was to assess the effect of fermentation duration and temperature on the efficacy and storage stability of freeze-dried cells of this bacterium.

Material and Methods

Preparation of pre-cultures of Pseudomonas fluorescens Pf153

Strain Pf153 was provided by the ETH Zürich and routinely cultivated on Tryptic Soy Agar plates (TSA; 30 g Tryptic Soy Broth (TSB, Difco, Germany) and 15 g agar-agar (Roth, Germany) in 1000 ml de-ionised water) at 25°C. To prepare the pre-culture inocula, cells were transferred by loop from TSA plates to 50-ml Erlenmeyer flasks containing 30 ml autoclaved King's medium B (KB; 20 g proteose peptone No. 3 (Difco), 10 ml glycerol (Merck, Germany), 1.5 g KH_2PO_4 (Merck), 1.5 $MgSO_4 \times 7$ H_2O (Merck) (King et al., 1954) and incubated at 28°C for 24 h at 150 rpm on a rotary shaker (Novotron, Infors, Switzerland).

Preparation of Botrytis cinerea inoculum

Botrytis cinerea provided by the JKI (Institute for Biological Control and Institute for Breeding Research on Fruit Crops: strains Bc111, 63444, 68731, 63451 and 62084), was grown on modified Czapek Dox agar (30 g skim milk powder (Saliter, Germany), 3 g $NaNO_3$, 1 g $K_2HPO_4 \times 3$ H_2O (Merck), 0.5 g $MgSO_4 \times 7$ H_2O, 0.5 g KCl, 0.01 g Iron (II)-sulfate $\times 7$ H_2O (Merck), and 18 g agar-agar brought up to 1000 ml with de-ionised water) in Petri dishes (diameter 9 cm) for 3 weeks at 20°C. Conidial suspensions were prepared by flooding the plates with 0.1% malt extract solution (Merck), gently scraping the mycelium with a spatula and filtering the resulting suspension through three layers of muslin in a glass funnel. The concentration was adjusted to 5×10^5 conidia ml^{-1} after counting the conidia with a Thoma haemocytometer.

Cultivation and preparation of P. fluorescens Pf153

Strain Pf153 was cultivated in 100-ml Erlenmeyer flasks or in a 4-l fermenter. The flasks and the fermenter were filled with 30 ml and 3.5 l KB medium, respectively, and autoclaved at 121°C for 20 min. After cooling, pre-cultures prepared as described above were added to the medium at a ratio of 1 : 100 and fermented at 20 or 28°C. At the end of the fermentation time, the cells were harvested and processed as described by Bisutti et al. (2015). Briefly, after repeated centrifugation, a cell suspension with optical density (OD_{595}) of 0.95 (± 0.02) was mixed at a ratio of 1 : 1 with a 20% (w/v) autoclaved lactose solution. Three ml of the mixture were transferred to sterile glass vials and freeze-dried with a freezing rate of 0.04–0.12°C min^{-1} to -40°C and followed by drying for 18 h at -20°C and 0.15 mbar. At the end of the process, the vials were sealed under 0.15 mbar and kept in the freezer (-18 °C) until use.

Determination of the number of viable cells

The freeze-dried product was made up to the original volume (3 ml) by adding sterile de-ionised water, and viable cells of Pfl53 in fresh cultures and the freeze-dried preparation were enumerated using serial dilutions and determination of MPN as described in Bisutti et al. (2015).

Determination of the fermentation parameters

The experiment was performed in a 4-l fermenter (ISF-100, Infors) at 20°C or at 28°C with 900 rpm agitation and $3 \cdot 70$ Nl min^{-1} air supply. Formation of foam was avoided by adding Antifoam (Roth) in automatic mode. Samples were taken every four hours. The cell density was measured as OD$_{595}$ using a SPECTRA Mini AP (Tecan, Switzerland). Samples were freeze-dried prior to addition to the cell suspension of a 20% (w/v) solution of lactose in ratio of 1 : 1. Freeze-drying followed the protocol described above. The experiment was repeated in five independent runs for fermentation at 28°C and three runs for fermentation at 20°C.

Bioassay against B. cinerea on detached Vicia faba leaves

The design of the bioassay was as described by Bisutti et al. (2015). Briefly, three compound leaves of *V. faba* cv 'con Amore' with four leaflets were placed together on a steel wire mesh with soaked filter paper underneath, in $20 \times 20 \times 5$ cm plexi glass boxes covered with a translucent lid (Gerda GmbH, Schwelm, Germany). Freeze-dried cells of Pfl53 produced in flasks as described above, were suspended with de-ionised water to a concentration of 5×10^8 MPN ml^{-1}. One ml of this suspension was sprayed onto each leaf surface using an air-brush sprayer; pathogen inoculation followed immediately by spraying one ml conidial suspension of *B. cinerea* on the upper side of the leaf. The boxes were then incubated at 20°C with a day/night cycle of 16/8 h. As control treatments water and a $0 \cdot 2\%$ (w/v) water suspension of Euparen (active ingredient: 50% Tolylfluanid, WG) were used. The disease severity was rated five days after inoculation by estimating the affected percent leaf area using the following rating scale: 0=No lesions, 1=0 to 1%, 2=2 to 5%, 3=6 to 10%, 4=11 to 25%, 5=26 to 50%, and 6=51 to 100%. A disease index (DI) was calculated as $DI = \frac{\Sigma(rating\ value \times number\ of\ leaves\ in\ the\ rating)}{total\ number\ of\ leaves \times highest\ rating\ value} \times 100$. The experiment included four boxes per treatment and was repeated three times.

Shelf life of freeze-dried cells of P. fluorescens Pfl53

The cells were produced in 100-ml Erlenmeyer flasks filled with 30 ml KB medium. The flasks were kept at 20 or 28°C on two identical rotary shakers at 150 rpm. Samples were then harvested at different times: 16 and 28 h for incubation at 20°C and 8 and 16 h for incubation at 28°C. Afterwards the samples were freeze-dried as described above. For measuring the storage stability, samples were incubated at 25°C in the dark. Bacterial viability was determined once a week by analysing two vials per sample. The experiment was time independent repeated four times.

Influence of incubation temperature on growth of **P. fluorescens** *Pf153in the fermenter and survival after freeze-drying*

The fermenter (Minifors, Infors) was kept at 20°C or at 28°C, 900 rpm agitation and 3·70 Nl min⁻¹ air supply. Foam formation was avoided by adding Antifoam (Roth) in automatic mode. Five hundred ml of the cell suspension was harvested in the middle and at the end of the log phase corresponding at 16 and 28 h for the fermenter at 20°C and 8 and 16 h for the fermenter at 28°C respectively. Thereafter, the samples were freeze-dried as described above and kept in the freezer at -18°C until use. The experiment was time independently repeated three times.

In vitro *tests of the influence of fermentation temperature and duration on the efficacy of* **P. fluorescens** *Pf153 against different* **B. cinerea** *strains*

The efficacy of freeze-dried cells of Pf153 produced in a fermenter at 20°C or 28°C, respectively, was tested *in vitro* against five strains of *B. cinerea*. The test was performed on two media differing in carbon source: 1/10 PDA and PDFA (PDFA sugar content of fructose and glucose 2 mg ml⁻¹ based on PDA (Hjeljord and Strømeg, 2004)). One hundred microliter each of a cell suspension of Pf153 prepared as described above, lactose solution (10% w/v) or Euparen (0·2% w/v) were pipetted in the middle of a 9 cm Petri plate and spread with a Drigalsky spatula. Immediately thereafter, the plates were inoculated with *B. cinerea* by placing a 5 mm mycelial plug of a culture from modified Czapek Dox agar in the centre of the plates. The plates were then placed in an incubator at 20°C with a 16/8 h night and day cycle. Radial mycelial growth was assessed at 4, 7, 10 and 16 days after inoculation (DAI) by measuring the diameter along two perpendiculars. A ranking based on the efficacy $\left[\frac{Mycelial\ growth\ with\ lactose-mycelial\ growth\ with\ Pf153}{Mycelial\ growth\ with\ lactose} \times 100\right]$ of the four Pf153 fermentations was made with the relative performance index (RPI) calculated for each fermentation temperature/duration combination. RPI for efficacy is dimensionless and between 0 and 100 because efficacy improves as disease rating decreases. The applied formula was $RPI = \left|\frac{x - x_{av}}{\sigma} - 2\right| \times 25$ where x is a single observation value for a strain, x_{av} is an average of all observations from all strains and σ is the standard deviation of all observations from all strains being ranked (Schisler and Slininger, 1997). Three Petri dishes were used for each fermentation temperature and *B. cinerea* combination. The test was repeated three times.

Statistical Analysis

Data were statistically analyzed with the software SAS System for Windows v9.3. Experiments with freeze-drying of Pf153 were analyzed using the generalized linear model. The Shapiro-Wilk test was applied for testing for normality. The homogeneity of variance was proven/checked by the Levene test ($P < 0·1$). For separation of the means, log10 transformed data were compared with the Student-Newman-Keuls test (SNK) ($P < 0·05$). Because in bioassays no homogeneity of variance or normality was achieved, the non-parametric test of Kruskal-Wallis was chosen. By the exact Methods in the

NPAR1WAY procedure (two-sided), the samples were compared pair wise (Wilcoxon, exact $P < 0.05$).

Results

Determination of the fermentation parameters

Growth profiles of Pf153 fermented at 20 and 28°C measured as OD of the cultures are shown in figure 1 left. The exponential phase of the cells cultured at 20°C was time delayed by about 8 h in comparison to the cells fermented at 28°C. The beginning of the stationary phase was reached after 16 h for cells fermented at 28°C and after 28 h for cells fermented at 20°C. The experiment was stopped after 24 h for the fermentation at 28°C. The samples were also freeze-dried. The survival rates after freeze-drying are shown in figure 1 (right). Freeze-dried cells from the beginning of the log phase cultured at 20 and 28°C showed a reduction in survival. Directly after the beginning of the log phase, the survival after freeze-drying increased with a maximum survival at the stationary phase for cells fermented at 28°C, while for cells grown at 20°C the values fluctuated between 80 and 100%.

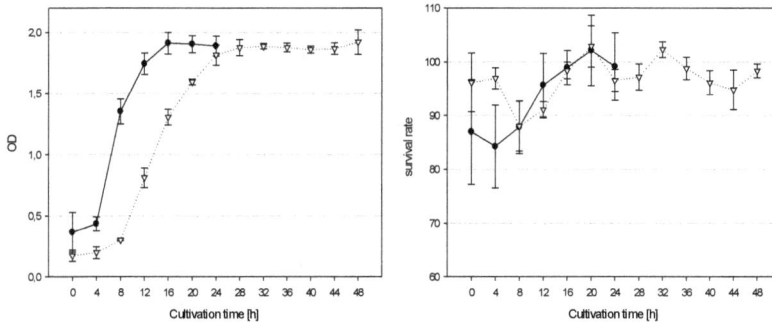

Figure 1. Growth curves for Pf153 grown at 20°C (triangle down) and 28°C (dots) in a fermenter measured as OD at 595-nm (left); Survival rates (calculated by dividing the log MPN after freeze-drying by the log MPN before freeze-drying, multiplied by 100). Values exceeding 100% are due to variability between measurements (right).

Bioassay against **B. cinerea** *on detached* **V. faba** *leaves*

Biocontrol efficacy of freeze-dried cells of Pf153 produced under different combinations of fermentation temperature and duration (20°C/16 h, 20°C/28 h, 28°C/8 h and 28°C/16 h) was assessed against *B. cinerea* on detached *V. faba* leaves (figure 2). All treatments reduced the disease severity significantly compared to the water control but not as effectively as the chemical Euparen (reduction respective to water control of 82·6%). The highest disease reduction was caused by cells harvested in the middle of the log phase

129

(56·5% for 28°C and 48·5% for 20°C grow temperature). Cells from the beginning of the stationary phase reduced disease at least by 37·9% (28°C/16 h).

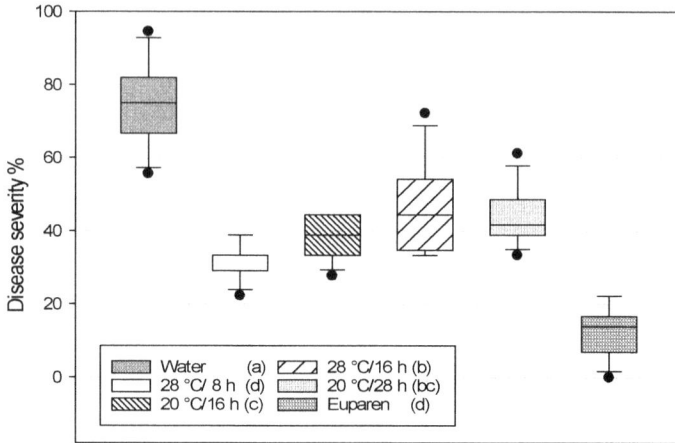

Figure 2. Disease severity of *B. cinerea* in a bioassay on detached leaves of *V. faba* with freeze-dried preparations from Pf153 cultured at different combinations of the fermentation parameters temperature and duration of culture. The box plots represent the minimum, 25%-75%, maximum and median values of disease severity.
Fermentation parameters in the legend followed by the same letter are not significantly different following SNK test (P<0·05).

Shelf life of freeze-dried cells of P. fluorescens *Pf153*

The samples of freeze-dried cells were stored at 25°C for 12 weeks and MPN ml^{-1} was assessed each week (figure 3). After 10 weeks the number of viable cells was reduced over time for all fermentation parameter combinations from about 10^9 to less than 10^6 MPN ml^{-1} for cells fermented at 28°C/8 h. The trend was for cells fermented at 20°C to show slightly better survival than those fermented at 28°C. Assessment at week 12 showed cell densities between 10^6 and 10^7 MPN ml^{-1}, with the exception of 28°C/8 h.

Influence of incubation temperature on growth of P. fluorescens *Pf153in the fermenter and survival after freeze-drying*

When cells were cultivated in the fermenter, the cell yield was highest for the cells fermented for 28 h at 20°C followed by those fermented for 16 h at 28°C. Freeze-drying of cells caused a reduction in the number of viable cells (table 1), statistically significant for the cells harvested at beginning of the stationary phase but not for those harvested in the middle of the log phase for both temperatures.

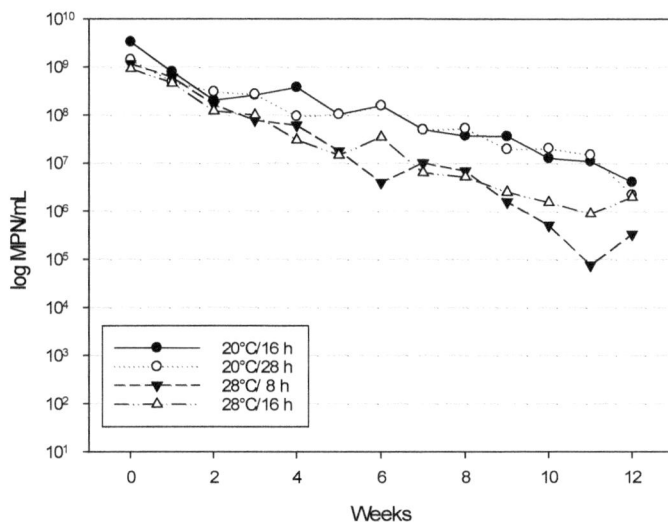

Temperature/ /Time	28°C/8 h	20°C/16 h	28°C/16 h	20°C/28 h
	y=-3·5124x+98·902	y=-2·1039x+96·209	y=-2·6846x+96.982	y=-2·0007x+98·111
	R^2=0·9467	R^2=0·9288	R^2=0·9374	R^2=0·9184

Figure 3. Influence of temperature and harvest time of Pf153 on viability (MPN ml^{-1}) during storage at 25°C for 12 weeks.

Table 1. Viability direct after harvest and before and after the freeze-drying process of Pf153 fermented at different temperature for different times

Fermentation parameters	Yield*	Before freeze-drying*	After freeze-drying*
28°C / 8 h	4·11 (± 1·51) x 10^9 (B)	9·62 (± 4·68) x 10^8 (a)	5·22 (± 2·38) x 10^8 (a)
20°C / 16 h	5·03 (± 0·62) x 10^9 (B)	2·34 (± 1·40) x 10^9 (a)	1·05 (± 0·70) x 10^9 (a)
28°C / 16 h	8·33 (± 2·16) x 10^9 (B)	1·57 (± 0·79) x 10^9 (a)	4·29 (± 1·80) x 10^8 (b)
20°C / 28 h	1·37 (± 0·76) x 10^{10} (A)	1·13 (± 0·34) x 10^9 (a)	5·64 (± 4·84) x 10^8 (b)

Note: *Viability is expressed as MPN ml^{-1} and the reported data are mean (± SD) of three independent experiments

Suspension prepared for freeze-drying had an OD of 0·95 ± 0·01

Means of the yield data followed by the same letter are not significantly different following SNK test (P<0·05)

Means of the fermentation parameters lines for before and after freeze-dried, followed by the same letter are not significantly different following SNK test (P<0·05)

In vitro *tests of the influence of fermentation temperature and duration on the efficacy of* P. **fluorescens** *Pf153 against different* B. **cinerea** *strains*

The assays for assessing the effect of co-inoculation of freeze-dried cells of Pf153 and *B. cinerea* were performed on the media PDFA and 1/10 PDA (figures S1 and S2). On control (lactose) plates, the speed of mycelial growth was similar for the *Botrytis* strains tested except for strain 62084 that grew considerably faster than the other strains. Compared to the lactose controls, the radial growth of all strains was significantly reduced on plates co-inoculated with Pf153. The reduction pattern was similar on both media. A strong, significant reduction was recorded for the chemical Euparen (figure S1, S2; table 2). The magnitude of the reducing effect of Pf153 on mycelial growth of the *B. cinerea* strains did not differ statistically between the different temperature/duration combinations used during fermentation of the bacterium (table 2). At first sight (figure S1, S2; table 2), no fermentation method showed an overall ability to reduce *B. cinerea* strains growth on both media. When calculating an RPI efficacy over all *B. cinerea* strains and media the cells from the fermentation 20°C/28 h showed the highest and those from the fermentation 28°C/16 h the lowest efficacy (table 3).

Table 2. Reduction of radial mycelial growth of five strains of *B. cinerea* on two different agar media in the presence of Pf153 or the chemical fungicide Euparen. Plates were inoculated with *B. cinerea* and freeze-dried cells of Pf153 from fermentations performed with different temperature/fermentation time combinations. Values are the percentage reduction of mycelial growth of *B. cinerea* relative to growth on control plates (lactose) determined after 10 days (strain 62084) or 16 days (all other strains) of co-cultivation.

	Bc111		63444		68731		63451		62084	
	PDFA	*1/10 PDA*	*PDFA*	*1/10 PDA*	*PDFA*	*1/10 PDA*	*PDFA*	*1/10 PDA*	*PDFA*	*1/10 PDA*
Lactose	0 (a)	0 (a)	0 (a)	0 (a)	0 (a)	0 (a)	0 (a)	0 (a)	0 (a)	0 (a)
28°C / 8 h	56 (b)	53 (b)	40 (a)	30 (a)	28 (b)	31 (ab)	38 (b)	37 (b)	62 (b)	70 (b)
20°C / 16 h	65 (b)	62 (b)	27 (a)	44 (a)	34 (b)	42 (b)	35 (b)	40 (b)	69 (b)	66 (b)
28°C / 16 h	61 (b)	60 (b)	22 (a)	19 (a)	25 (b)	33 (ab)	38 (b)	32 (b)	64 (b)	69 (b)
20°C / 28 h	65 (b)	61 (b)	31 (a)	38 (a)	38 (b)	37 (b)	46 (b)	43 (b)	71 (b)	71 (b)
Euparen	93 (c)	93 (c)	80 (b)	87 (b)	73 (c)	81 (c)	62 (c)	60 (c)	93 (c)	93 (c)

Statistics was made with the diametric mycelial growth values (means of three independent experiments), numbers within columns followed by the same letter are not significantly different according to the glimmix procedure
No statistical differences were found between mycelial growth on 1/10 PDA and PDFA

Table 3. Relative performance indices of Pf153 for reduction of radial mycelial growth of *B. cinerea* (compare table 2) on the media 1/10 PDA and PDFA

Fermentation parameters	1/10 PDA RPI$_{efficacy}$	PDFA RPI$_{efficacy}$	Combined RPI$_{efficacy}$
28°C / 8 h	47·45	47·31	48·49
20°C / 16 h	53·32	47·90	51·76
28°C / 16 h	44·91	42·93	45·01
20°C / 28 h	54·33	52·66	54·67

Discussion

A standard industrial method for mass production of bacterial and fungal cells is the liquid fermentation (Slininger and Shea-Wilbur, 1995). In case of bacteria or fungi used in biocontrol the aim is to obtain large amounts of viable micro-organism with high resistance to the formulation process, long shelf life and high efficacy (Ashofteh et al., 2009). In fact, when scaling up the production of a BCA, attention has to be paid to the efficacy because the process parameters yielding the best material for formulation are not necessarily optimal for producing biocontrol agents with a high efficacy (Angeli et al., 2016; Spadaro and Gullino, 2005). Previously, a freeze-drying protocol was established specifically for Pfl53 to achieve high viability after the process and during storage (Stephan et al., 2016). Subsequently, the influence of different culture parameters on the freeze-drying process were studied. During those studies, we found that Pfl53, fermented at 20°C for 16 h in KB medium in flasks and freeze-dried in the presence of lactose as lypoprotectant, showed increased efficacy against *B. cinerea* on broad bean detached leaves compared to the other tested fermentations parameters (Bisutti et al., 2015). In the present study, we therefore evaluate the influence of the fermentation parameters temperature and harvesting time on the efficacy against the pathogen *B. cinerea* of formulated cells of Pfl53 and cell survival during storage.

Growth curves were determined to evaluate the effect of temperature on growth kinetics by taking samples every 4 h and plotting the measured OD against time. Cells grown at 28°C reached the stationary phase faster than cells grown at 20°C, however, the yield in the mid-exponential phase at 20°C was 18·3% higher than at 28°C and at the beginning of the stationary phase this difference was even higher (39·2%). Increasing temperature increases cell growth but its influence on cell stability during formulation is not always positive. For example for lactic acid bacteria and other probiotics an increase of the temperature to 37°C influenced positively their growth however this temperature was not the best to obtain cells resistant to freeze-drying (Liu et al., 2014; Tanimomo et al., 2016). In the present study survival rates after freeze-drying changed slightly between experiments. Survival rate after freeze-drying for cells fermented at 20°C for 16 h was comparable with data reported by Bisutti et al. (2015) but for cells fermented at 28°C for 16 h the survival was lower than in that study. When considering viability after freeze-drying through all experiments (also from data not shown), the harvesting time seemed not to have an influence on the survival of Pfl53. Similar results were described by Saarela et al. (2005) for *Bifidobacterium animalis* ssp. *lactis*, however *Xanthomonas campestris* harvested during the early or late stationary phases showed a significantly higher survival than log phase harvested cells (Jackson et al., 1998). In storage experiments, some *Pseudomonas* failed to survive storage, however others lived up to 20 years. These different survival behaviours are attributed to the storage temperature: *Pseudomonas* survive better when stored at temperatures lower than 8°C as freeze-dried cells (Miyamoto-Shinohara et al., 2006; Palmfeldt et al., 2003). These better storage at standard refrigeration temperature was also reported for other Gram-negative bacteria like *Xanthomonas campestris* (Jackson et al., 1998) and probiotic bacteria (Saarela et al., 2005; Tanimomo et al., 2016). Storage below freezing temperatures could be one major obstacle to the large-scale use of *Pseudomonas* formulations (Slininger et al., 1996; Stockwell and Stack, 2007) because not all consumers can afford equipment to keep the product at low temperature. In our trials we stored the vacuum sealed vials containing the freeze-dried Pf153 at 25°C for up to 12 weeks and tested each week the viability by MPN ml^{-1}. The survival was reduced for all culture conditions while cells fermented at 28°C showed a faster reduction than the 20°C cultivated cells, especially after the fifth week of storage. At week nine, the survival rate of cells fermented at 28°C was as low as the 20°C fermented at week 12. For both temperatures, cells harvested at the beginning stationary phase showed better survival. For the commercially available bioproducts containing *Pseudomonas*, the reported storage temperature from -18 °C to 8 °C resulted in a shelf life ranging from eight months to weeks. At room temperature the product containing *P. chlororaphis* is warranted for three weeks (Berninger et al., 2018). The formulate of *Arthrobacter chlorophenolicus* A6, which can be used for bioremediation, was stable and viability for three months when stored at 4°C, without efficacy reduction compared to the fresh cells (Bjerketorp et al., 2018). Pf153 was stable after eight weeks at 25°C, with still at least $5 \cdot 2 \times 10^7$ MPN ml^{-1}. The shelf life of freeze-dried Pf153 is, therefore, in the given interval for other *Pseudomonas*. Various mechanisms, like culture physiology and environment, have been described to influence cell survival. Multiple cell protection mechanisms that occur at different times during culturing can result in differences in drying and storing survival characteristics of cells of different age (Slininger et al., 1996). This was shown for *P. putida* where culture conditions and physiological state notably influenced the bacterial tolerance to the freeze-drying process (Muñoz-Rojas et al., 2006). The authors proposed that this may be related to changes in membrane properties during growth. Temperature, pH, culture medium composition and the transition to the stationary phase alter the fatty acid content of the cell membrane of *P. fluorescens*. These modifications change the membrane viscosity and influence cellular functions including cell growth and stability (Fouchard et al., 2005). We presume that the different storage behaviour is due to changes in the phospholipid fatty acids of the cells with increased ratio of unsaturated to saturated fatty acids making the cell more resistant to the freeze-drying conditions used and consequently for storage. An increase in unsaturated fatty acids in the cell membrane maintains fluidity and stability and decreases membrane leakage during rehydration (Liu et al., 2014).

The influence of the different fermentation procedures on the efficacy of Pf153 was tested against five different *B. cinerea* strains *in vitro* with two media differing in sugar composition. The medium PFLA was proposed by Hjeljord and Strømeg (2004) when investigating BCAs against *B. cinerea* on strawberry blossoms. The authors developed the medium on the basis of the extract composition of the blossoms (Hjeljord and Strømeg, 2004). In our case none of the temperature/duration combinations used with Pf153 was better against all *B. cinerea* strains and no statistical differences were found between the media used in the test. When considering RPI the cells grown at 20°C show better performance than the ones grown at 28°C despite the age. The five *Botrytis* strains showed different growth rates on the two media and sensitivity to the chemical control Euparen. Our results were similar to the results of Bardin et al. (2013) who showed that the efficacy of a BCA can vary, in certain conditions, depending on the *B. cinerea* strain. In test with *Microdochium dimerum* and Serenade Max$^{®}$ (*Bacillus subtilis* QST713) against different

B. cinerea strains on tomato and lettuce plants, the protection provided by the BCAs was significantly influenced by the pathogen strain and its aggressiveness level (Bardin et al., 2013a; Bardin et al., 2013b). In other studies, variations in the sensitivity to antimicrobial compounds produced by BCAs to pathogens isolates was shown: resistance or tolerance to 2,4-diacetylphloroglucinol for *G. graminis* or *Fusarium oxsporum* and a wide range of sensitivity against pyrrolnitrin for *B. cinerea* (Bardin et al., 2015). *B. cinerea* is a weak plant pathogen that can antagonizes other micro-organisms on plant surfaces. For example, it produces botrydial and botrycine, active against Gram-positive bacteria and fungi or yeast respectively (Blakeman and Fokkema, 1982). *Pseudomonas* can lyse fungal cells and degrade metabolites produced by pathogens that induce pathogenesis (Bhattacharjee and Dey, 2014). Its ability to suppress growth and to modify hyphal structure was shown for *B. cinerea* (Barka et al., 2002; Blakeman and Fokkema, 1982) and *Drechslera dictyoides* (Blakeman and Fokkema, 1982). On the other hand, *Pseudomonas* can stimulate germination and infection of certain leaf-infecting pathogens on leaf and fruit surfaces (Blakeman and Fokkema, 1982). It is reported that many phytopathogenic fungi are sensitive to volatile compounds produced by bacteria. Hydrogen cyanide, for example, is thought to be involved in root pathogen suppression (Pal and Gardener, 2006) and fluorescent pseudomonads biocontrol activity and hydrogen cyanide production ability are hypothesized to have a close relationship with virulence (Jain and Das, 2016). Pfl53 synthesizes different antifungal compounds including hydrogen cyanide and an extracellular protease (Fuchs et al., 2000).

The performance of a biopesticide is enhanced by fermentation and formulation which are closely linked (Hynes and Boyetchko, 2006). When optimizing the production of a *Pseudomonas*, the growth conditions need to be considered carefully (Fuchs et al., 2000). For quality control it is also important to consider several target pathogen strains to obtain a better representation of the pathogen population (Bardin et al., 2013b) in the field. The improvement of a BCA during production allows the enhanced survival and activity under differing environmental conditions improving its potential as commercial product (Teixidó et al., 2005). Our experiments clearly demonstrate that by manipulating fermentation parameters it is possible to increase the storage survival of Pfl53. In our case fermentation at 20°C for 28 h showed the highest yield and storage survival at 25°C. Freeze-dried cells produced under these conditions showed the best performance in *in vitro* tests against different *Botrytis* strains when considering the RPI but not when tested on detached leaves. This shows the importance of the efficacy testing system comprising pathogen diversity. However, only greenhouse and field trials could demonstrate if performance is also achievable in field conditions and if culture conditions have a stronger influence on field efficacy. To achieve a better quality and reliable efficacy of BCAs, optimization of production and formulation protocols are unavoidable.

Acknowledgements

This research was supported by SafeCrop Centre, funded by Fondo per la Ricerca, Autonomous Province of Trento. We have to thank Eckhard Koch (JKI, Germany) for proofreading the manuscript.

References

Angeli, D., Saharan, K., Segarra, G., Sicher, C., and Pertot, I. (2016). Production of *Ampelomyces quisqualis* conidia in submerged fermentation and improvements in the formulation for increased shelf-life. *Crop Protection* **97**(July):135-144.

Ashofteh, F., Ahmadzadeh, M., and Fallahzadeh-Mamaghani, V. (2009). Effect of mineral components of the medium used to grow biocontrol strain UTPF61 of *Pseudomonas fluorescens* on its antagonistic activity against *Sclerotinia* wilt of sunflower and its survival during and after formulation process. *Journal of Plant Pathology* **91**(3):607-613.

Bardin, M., Ajouz, S., Comby, M., Lopez-Ferber, M., Graillot, B., Siegwart, M., and Nicot, P. C. (2015). Is the efficacy of biological control against plant diseases likely to be more durable than that of chemical pesticides? *Frontiers in Plant Science* **6:566**:1-13.

Bardin, M., Comby, M., Lenaerts, R., and Nicot, P. C. (2013a). Diversity in susceptibility of *Botrytis cinerea* to biocontrol products inducing plant defence mechanisms. *IOBC-WPRS Bull* **88**:45-49.

Bardin, M., Comby, M., Troulet, C., and Nicot, P. C. (2013b). Relationship between the aggressiveness of *Botrytis cinerea* on tomato and the efficacy of biocontrol. *IOBC-WPRS Bull* **86**:163-168.

Barka, E. A., Gognies, S., Nowak, J., Audran, J.-C., and Belarbi, A. (2002). Inhibitory effect of endophyte bacteria on *Botrytis cinerea* and its influence to promote the grapevine growth. *Biological Control* **24**(2):135-142.

Bashan, Y., de-Bashan, L. E., Prabhu, S. R., and Hernandez, J.-P. (2014). Advances in plant growth-promoting bacterial inoculant technology: formulations and practical perspectives (1998-2013). *Plant and Soil* **378**(1-2):1-33.

Berninger, T., González López, Ó., Bejarano, A., Preininger, C., and Sessitsch, A. (2018). Maintenance and assessment of cell viability in formulation of non-sporulating bacterial inoculants. *Microbial Biotechnology* **11**(2):277-301.

Bhattacharjee, R., and Dey, U. (2014). An overview of fungal and bacterial biopesticides to control plant pathogens/diseases. *African Journal of Microbiology Research* **8**(17):1749-1762.

Bisutti, I. L., Hirt, K., and Stephan, D. (2015). Influence of different growth conditions on the survival and the efficacy of freeze-dried *Pseudomonas fluorescens* strain Pf153. *Biocontrol Science and Technology* **25**(11):1269-1284.

Bjerketorp, J., Röling, W. F., Feng, X.-M., Garcia, A. H., Heipieper, H. J., and Håkansson, S. (2018). Formulation and stabilization of an *Arthrobacter* strain with good storage stability and 4-chlorophenol-degradation activity for bioremediation. *Applied Microbiology and Biotechnology* **102**:2031-2040.

Blakeman, J. P., and Fokkema, N. (1982). Potential for biological control of plant diseases on the phylloplane. *Annual Review of Phytopathology* **20**(1):167-190.

Corrêa, E., Sutton, J., and Bettiol, W. (2015). Formulation of *Pseudomonas chlororaphis* strains for improved shelf life. *Biological Control* **80**:50-55.

Fouchard, S., Abdellaoui-Maâne, Z., Boulanger, A., Llopiz, P., and Neunlist, S. (2005). Influence of growth conditions on *Pseudomonas fluorescens* strains: A link between metabolite production and the PLFA profile. *Fems Microbiology Letters* **251**(2):211-218.

Fuchs, J. G., Moënne-Loccoz, Y., and Défago, G. (2000). The laboratory medium used to grow biocontrol *Pseudomonas* sp Pf153 influences its subsequent ability to protect cucumber from black root rot. *Soil Biology & Biochemistry* **32**(3):421-424.

Gao, X. N., He, Q. R., Jiang, Y., and Huang, L. L. (2016). Optimization of nutrient and fermentation parameters for antifungal activity by *Streptomyces lavendulae* Xjy and its biocontrol efficacies against *Fulvia fulva* and *Botryosphaeria dothidea*. *Journal of Phytopathology* **164**(3):155-165.

He, L., Xu, Y.-Q., and Zhang, X.-H. (2008). Medium factor optimization and fermentation kinetics for phenazine-1-carboxylic acid production by *Pseudomonas* sp M18G. *Biotechnology and Bioengineering* **100**(2):250-259.

Hjeljord, L. G., and Strømeg, G. M. (2004). Biological control of *Botrytis*: searching for a realistic screen. *Antalya, Turkey*:63.

Hynes, R. K., and Boyetchko, S. M. (2006). Research initiatives in the art and science of biopesticide formulations. *Soil Biology & Biochemistry* **38**(4):845-849.

Jackson, M., Frymier, J., Wilkinson, B., Zorner, P., and Evans, S. (1998). Growth requirements for production of stable cells of the bioherbicidal bacterium *Xanthomonas campestris*. *Journal of Industrial Microbiology and Biotechnology* **21**(4-5):237-241.

Jain, A., and Das, S. (2016). Insight into the interaction between plants and associated fluorescent *Pseudomonas* spp. *International Journal of Agronomy* **2016**.

King, E. O., Ward, M. K., and Raney, D. E. (1954). Two simple media for the demostration of pyocyanin and fluorescin. *Journal of Laboratory and Clinical Medicine* **44**(2):301-307.

Lee, J. P., Lee, S. W., Kim, C. S., Son, J. H., Song, J. H., Lee, K. Y., Kim, H. J., Jung, S. J., and Moon, B. J. (2006). Evaluation of formulations of *Bacillus licheniformis* for the biological control of tomato gray mold caused by *Botrytis cinerea*. *Biological Control* **37**(3):329-337.

Liu, X., Hou, C., Zhang, J., Zeng, X., and Qiao, S. (2014). Fermentation conditions influence the fatty acid composition of the membranes of *Lactobacillus reuteri* I5007 and its survival following freeze-drying. *Letters in applied microbiology* **59**(4):398-403.

Manikandan, R., Saravanakumar, D., Rajendran, L., Raguchander, T., and Samiyappan, R. (2010). Standardization of liquid formulation of *Pseudomonas fluorescens* Pf1 for its efficacy against *Fusarium* wilt of tomato. *Biological Control* **54**(2):83-89.

Miyamoto-Shinohara, Y., Sukenobe, J., Imaizumi, T., and Nakahara, T. (2006). Survival curves for microbial species stored by freeze-drying. *Cryobiology* **52**(1):27-32.

Mputu, K. J.-N. (2014). Impact du séchage sur la viabilité de *Pseudomonas fluorescens* (synthèse bibliographique). *Biotechnologie, Agronomie, Société et Environnement= Biotechnology, Agronomy, Society and Environment [= BASE]* **18**(1):134-141.

Muñoz-Rojas, J., Bernal, P., Duque, E., Godoy, P., Segura, A., and Ramos, J. L. (2006). Involvement of cyclopropane fatty acids in the response of *Pseudomonas putida* KT2440 to freeze-drying. *Applied and Environmental Microbiology* **72**(1):472-477.

Pal, K. K., and Gardener, B. M. (2006). Biological control of plant pathogens. *The plant health instructor* **2**:1117-1142.

Palmfeldt, J., Rådström, P., and Hahn-Hägerdal, B. (2003). Optimisation of initial cell concentration enhances freeze-drying tolerance of *Pseudomonas chlororaphis*. *Cryobiology* **47**(1):21-29.

Paulitz, T. C., and Belanger, R. R. (2001). Biological control in greenhouse systems. *Annual Review of Phytopathology* **39**:103-133.

Saarela, M., Virkajärvi, I., Alakomi, H. L., Mattila-Sandholm, T., Vaari, A., Suomalainen, T., and Mättö, J. (2005). Influence of fermentation time, cryoprotectant and neutralization of cell concentrate on freeze-drying survival, storage stability, and acid and bile exposure of *Bifidobacterium animalis* ssp *lactis* cells produced without milk-based ingredients. *Journal of Applied Microbiology* **99**(6):1330-1339.

Schisler, D. A., and Slininger, P. J. (1997). Microbial selection strategies that enhance the likelihood of developing commercial biological control products. *Journal of Industrial Microbiology & Biotechnology* **19**(3):172-179.

Slininger, P. J., and Shea-Wilbur, M. A. (1995). Liquid-culture pH, temperature, and carbon (not nitrogen) source regulate phenazine productivity of take-all biocontrol agent *Pseudomonas fluorescens* 2-79. *Applied Microbiology and Biotechnology* **43**(5):794-800.

Slininger, P. J., VanCauwenberge, J. E., Bothast, R. J., Weller, D. M., Thomashow, L. S., and Cook, R. J. (1996). Effect of growth culture physiological state, metabolites, and formulation on the viability, phytotoxicity, and efficacy of the take-all biocontrol agent *Pseudomonas fluorescens* 2-79 stored encapsulated on wheat seeds. *Applied Microbiology and Biotechnology* **45**(3):391-398.

Spadaro, D., and Gullino, M. L. (2005). Improving the efficacy of biocontrol agents against soilborne pathogens. *Crop Protection* **24**(7):601-613.

Stephan, D., Da Silva, A. P. M., and Bisutti, I. L. (2016). Optimization of a freeze-drying process for the biocontrol agent *Pseudomonas* spp. and its influence on viability, storability and efficacy. *Biological Control* **94**:74-81.

Stockwell, V. O., and Stack, J. P. (2007). Using *Pseudomonas* spp. for integrated biological control. *Phytopathology* **97**(2):244-249.

Tanimomo, J., Delcenserie, V., Taminiau, B., Daube, G., Saint-Hubert, C., and Durieux, A. (2016). Growth and freeze-drying optimization of *Bifidobacterium crudilactis*. *Food and Nutrition Sciences* **7**(7):616-626.

Teixidó, N., Cañamás, T. P., Usall, J., Torres, R., Magan, N., and Viñas, I. (2005). Accumulation of the compatible solutes, glycine-betaine and ectoine, in osmotic stress adaptation and heat shock cross-protection in the biocontrol agent *Pantoea agglomerans* CPA-2. *Letters in Applied Microbiology* **41**(3):248-252.

SUPPLEMENTS

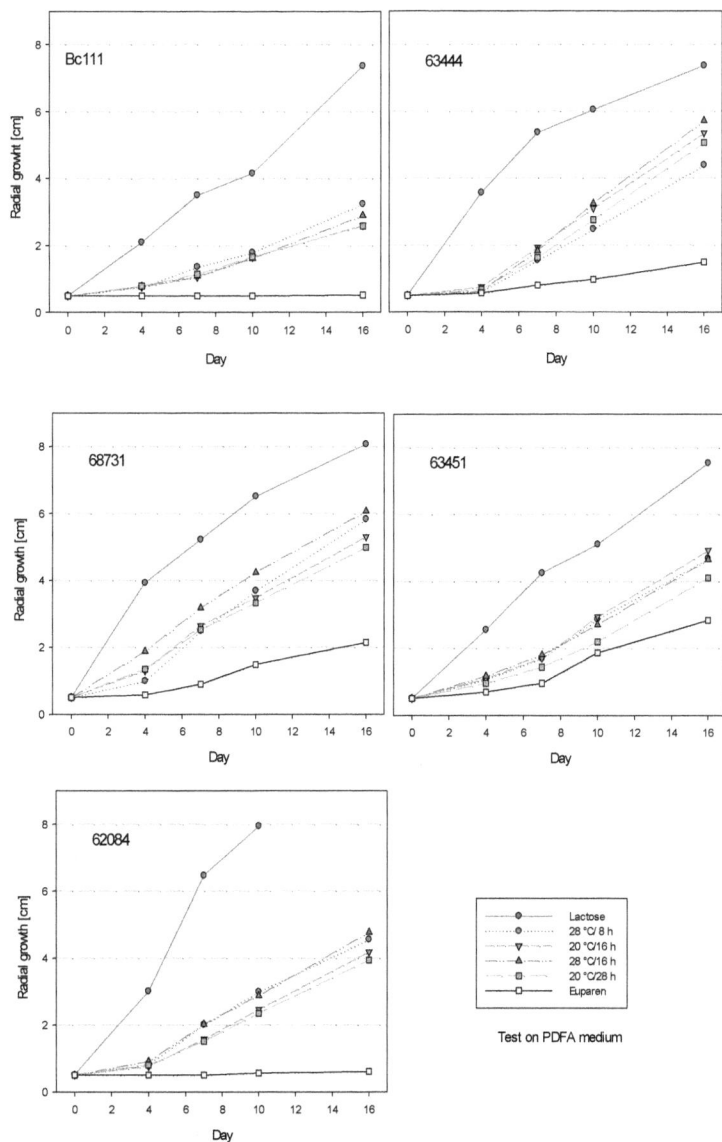

Figure S1. Mycelial growth of five different *Botrytis cinerea* on PDFA medium in the presence of formulated Pf153 cultivated at 20 or 28°C and harvested at different growth phases (mid-exponential and at the beginning of the stationary phase).

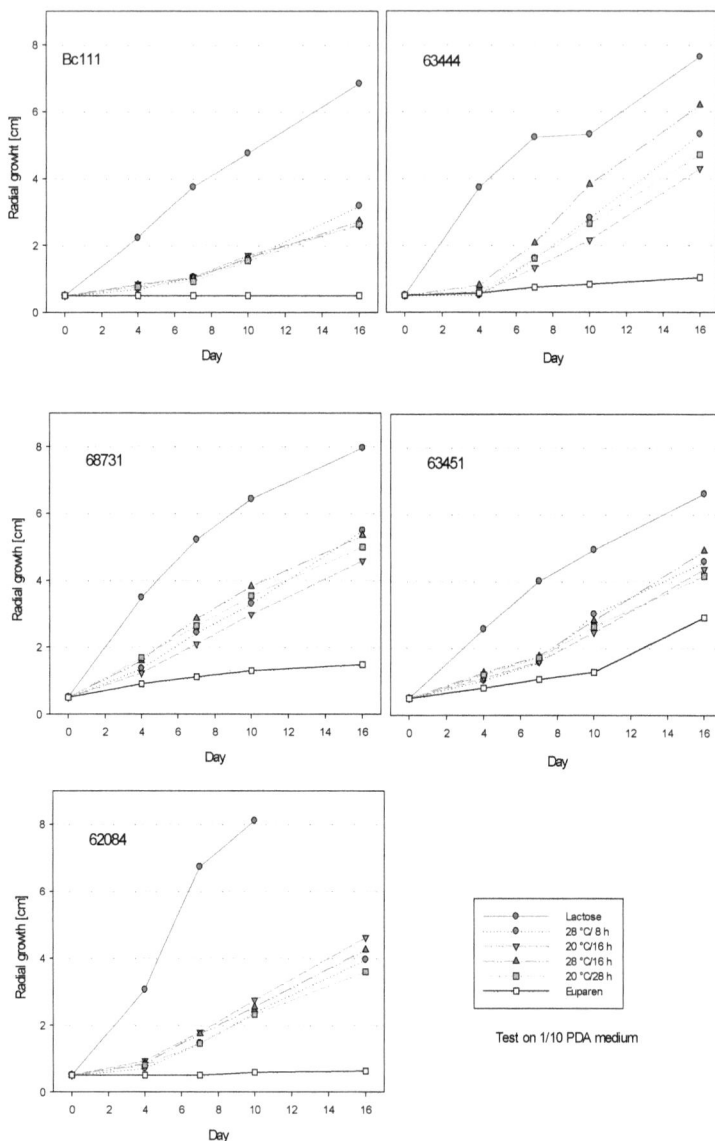

Figure S2. Mycelial growth of five different *Botrytis cinerea* on 1/10 PDA medium in the presence of formulated Pf153 cultivated at 20 or 28°C and harvested at different growth phases (mid-exponential and at the beginning of the stationary phase).

Field assessment on the influence of RhizoVital® 42 fl. and Trichostar® on strawberries in the presence of soil-borne diseases (2017).

Bisutti, I.L., Pelz, J., Büttner, C. and Stephan, D.

Crop Protection 96: 195-203

http://dx.doi.org/10.1016/j.cropro.2017.02.004

Crop Protection 96 (2017) 195-204

Contents lists available at ScienceDirect

Crop Protection

journal homepage: www.elsevier.com/locate/cropro

Field assessment on the influence of RhizoVital® 42 fl. and Trichostar® on strawberries in the presence of soil-borne diseases

I.L. Bisutti [a, b], J. Pelz [a], C. Büttner [b], D. Stephan [a, *]

[a] Julius Kühn-Institut, Federal Research Centre for Cultivated Plants, Institute for Biological Control, Germany
[b] Humboldt-Universität zu Berlin, Faculty of Life Science, Section Phytomedicine, Germany

ARTICLE INFO

Article history:
Received 1 October 2016
Received in revised form
26 January 2017
Accepted 5 February 2017
Available online 2 March 2017

Keywords:
Verticillium wilt
Bacillus amyloliquefaciens
Trichoderma harzianum
Microsclerotia

ABSTRACT

Strawberries are one of the most popular fruits in Germany. However, an expansion of the cultivation area, especially for organic strawberry production, is limited due to the high yield losses caused by diseases and arthropod pests. Therefore, the two soil strengthener RhizoVital® 42 fl. and Trichostar were evaluated for their efficacy in stabilizing strawberry yields in the presence of soil-borne disease

First, an integration of antagonists in an IPM strategy was proven by testing their compatibility with synthetic pesticides. When mixed for a maximum of four hours only few fungicides reduced germination of the antagonistic fungus contained in Trichostar®. The germination of the bacterial antagonist (Rhizovital® 42 fl.) was not influenced by any tested synthetic pesticide used in this study. Afterwards, the microbial based products were tested in two commercial strawberry fields in the Rheine-Main area, which Verticillium-wilt was found. In one field, with just one treatment of Trichostar® in 2013, the yield increase in 2013 and 2014 was 8 and 9%, respectively. On the second field, RhizoVital® 42 fl. was applied both years and the increase was about 6% each year.

Our results demonstrate that beneficial micro-organisms can promote health and can stabilize increase yield of strawberries under field conditions. Integration of these organisms in IPM strategy seems to be possible.

© 2017 Elsevier Ltd. All rights reserved.

1. Introduction

The worldwide production of strawberry (Fragaria x ananassa) in 2013 was more than 7.7 million tons. Europe produced 19.2% of the worldwide strawberry production and about 10% of the European production was harvested in Germany (FAO, 2016). In 2015, out of 172,588 tons of total strawberry production, only 1.74% were organically grown strawberries (Bundesamt, 2016). Major concerns for strawberry producers throughout the world are soil-borne pathogens and weeds. In commercial strawberry production yield losses caused by fungal diseases and nematodes of up to 20—30% have been estimated (Conti et al., 2014). The potential of biocontrol agents allows progress in plant protection and qualities demands while reducing applications of chemical pesticides.

Strawberries are affected by various fruit (e.g. Botrytis cinerea and Phytophthora cactorum), foliar (e.g. Podosphaera aphanis, Mycosphaerella fragariae and Diplocarpon earliana) and crown and

root (e.g. Phytophthora cactorum, Phytophthora fragariae) disease Diseases on the fruit reduce yield directly, whereas plant disease on the plant reduce yield indirectly by weakening the plant, causir it to die. In fact, diverse soil pathogens like Phytophthora sp Rhizoctonia, black root rot originating from pathogen complex, a responsible for dramatic yield losses (Berg, 2007; Martin and Bu 2002) and increase soil infestation. In particular, since the midd of the 1990, Verticillium wilt has been causing considerable damag to strawberries in Germany (Neubauer and Heitmann, 2011). Ve ticillium wilt, mainly caused by V. dahliae, occurs throughout th temperate zones of the world and is favoured by environment stresses like sudden temperature changes (Maas, 1998) as well dry and warm summers (Harris and Yang, 1996). V. dahliae produc microsclerotia (MS) in the plant tissue which remaining in soil aft degradation of the plant. MS are the principal source of inoculu for the disease (Harris and Yang, 1996) and can be disseminate with wind and water (Maas, 1998). The wilt is influenced by cultiv susceptibility, weather, soil type (Harris and Yang, 1996), temper ature and humidity and indirectly by the soil micro flora and fau influencing the MS germination (Neubauer and Heitmann, 2011)

* Corresponding author.
E-mail address: dietrich.stephan@julius-kuehn.de (D. Stephan).

http://dx.doi.org/10.1016/j.cropro.2017.02.004
0261-2194/© 2017 Elsevier Ltd. All rights reserved.

142

Since Montreal Protocol on substances depleting the ozone layer is executed, the use of the very effective fumigant methyl bromide to disinfect the soil from pests is no longer allowed. The replacement with other chemicals was only temporary (Colla et al., 2012). Since the beginning of 2014, the European Directive 2009/128/EC on sustainable use of pesticides requires the application of the general principles of the Integrated Pest Management (IPM) in which non-chemical methods must be preferred and pesticides should have the least possible impact on non-target organism and the environment (Colla et al., 2012). Biological and cultural alternative strategies for disease control such as crop rotation, biofumigation, solarization, use of tolerant or resistant cultivars and lean planting stocks (Martin and Bull, 2002) are of increasing interest. Another alternative to the use of pesticides protecting roots against fungal pathogens are antagonistic rhizobacteria (Kurze et al., 2001) and fungi (Elad, 2000; Howell, 2003). The application of this biological strategy should influence the rhizosphere to optimize plant growth also by enhancing beneficial and reducing deleterious micro-organisms (Martin and Bull, 2002).

Bacteria can be beneficial for plant growth, yield and crop quality by antagonistic activity against detrimental microbes (Esitken et al., 2010). For example isolates of the genus Bacillus are often reported as growth-promoting bacteria (Nehra and Choudhary, 2015) but also as antagonists of different pathogens like Rhizoctonia solani on lettuce (Chowdhury et al., 2013) or Colletotrichum spp. on strawberries (Nam et al., 2014). Bacillus spp. exert their beneficial activity by producing antifungal and antibacterial metabolites and they can induce systemic plant resistance (Chowdhury et al., 2013). Beside plant growth promoting bacteria a number of fungi have been reported as antagonists of plant pathogens. For example Trichoderma species have been known to be antagonistic to Botrytis on grapes (Elad, 2000) and strawberries (Freeman et al., 2004; Levy et al., 2015) and against diverse soilborne pathogens (Kredics et al., 2003) like Rhizoctonia on mungbeans (Dubey et al., 2011) and strawberries (Elad et al., 1981), Phytophthora cactorum (Porras et al., 2007) and Colletotrichum on strawberries (Freeman et al., 2004). In particular T. harzianum Rifai has been identified as a very promising biocontrol agent because of the antagonistic effect on a wide range of plant pathogenic fungi (Saxena et al., 2014).

In this study, the two commercial products RhizoVital® 42 fl. and Trichostar®, generally used as soil strengtheners, were tested to assess their ability to protect strawberry plants against soil-borne pathogens, their influence on yield and their capacity to manage V. dahliae MS. Additionally the possibility of integration in a IPM strategy was also considered.

2. Material and methods

2.1. Plant and microbial material

Commercial strawberry plants (cv. "Honeoye", Kraege Beerenpflanzen GmbH & Co. KG, Germany) were delivered as frigo (cold stored bare rooted) and green (fresh harvested runner) plants and directly planted.

RhizoVital® 42 fl. (Bacillus amyloliquefaciens FZB42; Abitep GmbH, Germany) was prepared in tap water with a concentration of 2.5 × 10⁷ spores/ml and used directly. Trichostar® (Trichoderma harzianum T58; Gerlach GmbH & Co. KG, Germany) was activated in lukewarm tap water for 4 h before use in a concentration of 10⁵ spores/ml. Both products were prepared as advised by the producers. Their combination (here after referred to as Mixture) was prepared in the reported concentration with Trichostar® activated for 4 h and RhizoVital® 42 fl. added just before application.

2.2. Laboratory experiments on the compatibility of biological with chemical pesticides

The 14 pesticides used in conventional strawberry cultivation (Table 1) were mixed with the biological products to test their influence on the viability of the active microbial ingredients. For this purpose, concentrations of the chemical pesticides were made with tap water before mixing with the microbial products. The end concentration of the pesticides is given in Table 1 and the end concentration of RhizoVital® 42 fl. was 0.1% (v/v) and of Trichostar® 10% (v/v). Directly after mixing and after 4 h incubation at room temperature, the germination of the spores was determined. In the case of RhizoVital® 42 fl. 40 µl of the mixture were plated on Tryptic Soy Agar (30 g Tryptic Soy Broth (Difco, Germany) and 15 g agar-agar (Roth, Germany) in 1000 ml de-ionised water) by means of a spiral plater (Model C, Spiral Systems, Inc. Ohio USA). After 24 h incubation of the plates at 25 °C the number of colony forming units (CFU) was counted and the number of CFU per ml was calculated. In the case of Trichostar® 100 µl of the suspension was distributed on Malt Peptone Agar supplemented with antibiotics (30.0 g malt extract (Merck), 5.0 g soy peptone (Merck), 15 g agar-agar (Roth), 30.0 mg streptomycin sulfate (Sigma, Germany), 50.0 mg chloramphenicol (Roth) and 50.0 mg benomyl (Aldrich, Germany) in 1000 ml de-ionised water) with a Drigalsky spatula. After 24 h of incubation at 25 °C, the germination rate of 100 spores randomly checked under the microscope at different areas of the plate was determined. The entire experiment was repeated thrice with three replicates per repetition.

2.3. Field trials

Field trials were conducted at two different locations in south Hesse, Germany, known to have V. dahliae infections. Commercial field 1 (sand 37.5%, silt and clay 62.2%; pH 6.75; without artificial irrigation; previous crop: strawberry) was located in Bergen-Enkheim and commercial field 2 in Bischofsheim (sand 80.4%, silt and clay 19.6%; pH 8.21; with artificial irrigation; previous crop: green manure with Raphanus sativus). Planting followed the farm standards with matted rows in Bergen-Enkheim and raised beds in Bischofsheim. Before planting, the roots of 25 plants were dipped for 15 min in 2 l of the antagonist suspension or in 2 l tap water used as control. There were an untreated control and three treatments: RhizoVital® 42 fl., Trichostar®, and the Mixture. Strawberries were planted in rows distanced 90 cm with a distance of 25 cm between the plants. Field 1 was planted on 09 May 2012 with frigo plants and field 2 on 23 August 2012 with green plants. After planting the plots were tilled by the farmers. Two months after planting and again in the spring of 2013 the 25 strawberry plants (six meter row) were treated with 5 l treatment suspensions of antagonists (corresponding to 0.926 l/m²). Field 2 was additionally treated in spring 2014. For each treatment four repetitions in a random block design were set up. Fruit yield was assessed by picking all ripe and healthy fruits. In 2013 the harvest started at the beginning of June, which was abnormal late. In 2014 the harvesting time started two weeks before 2013. In field 1 four pickings per season were made by seasonal harvest crew. Field 2 was picked six times in 2013 and ten times in 2014 by student assistants. In 2013 and 2014 the extent of plant damage was visually rated at the end of the harvesting season. The applied scale was from zero to two with 0 = plant dead, 1 = plant diseased (wilting of the older leaves) and 2 = plant symptomless as described by Vestberg et al. (2008). Field 1 was not visually rated in 2014 due to the high weed infestation.

Additionally, soil samples were taken before and after planting to determine the number of MS. The samples were collected with a soil probe (2 cm diameter) by drilling a 20 cm deep core. Before

Table 1
Commercial chemical products tested for their compatibility with RhizoVital® 42 fl. (based on *Bacillus amyloliquefaciens* FZB42) and Trichostar® (based on *Trichoderma harzianum* T58).

Active ingredient/chemical product	Concentration %	Trade name	target (Pest/diseases controlled)
Kresoxim-methyl	0.015	Discus®	Powdery mildew
Trifloxystrobin	0.015	Flint®	Powdery mildew, leaf scorch and spot
Azoxystrobin	0.05	Ortiva®	*Colletotrichum*, additional effect on mildew, *Botrytis*
Boscalid Pyraclostrobine	0.09	Signum®	Grey mould, *Gnomonia*, leaf scorch and spot
Quinoxyfen	0.025	Fortress® 250	Powdery mildew
Cyprodinil Fludioxonil	0.05	Switch®	Grey mould
Difenoconazole	0.02	Score®	Leaf scorch and spot, *Gnomonia*
Penconazole	0.025	Topas®	Powdery mildew
Fenhexamid	0.10	Teldor®	Grey mould
Fosetyl-Al	0.50–1	Aliette®	Crown rot (*P. cactorum*) – red stele root rot (*P. fragariae*)
Copper oxychloride	0.05	Funguran®	Angular leaf spot
Thiacloprid	0.0125	Calypso®	Aphids, reduction of strawberry bud weevil
Abamectin	0.0625	Vertimec®	Spider mites, strawberry mite, thrips (GW)
Pendimethalin	0.167	Stomp® Aqua	Annual dicots weeds (except cleavers, chamomile species, gallant soldier, common groundsel)

planting 40 samples per field were taken on a grid pattern; after planting four samples per treatment and repetition were taken between plants in the rows. The soil samples were then pooled, mixed and allowed to air dry at 20 °C for 2 weeks. Afterwards they were sieved (2 mm) and stored in plastic bags in a cool and dark room until analysis. The MS density was determined by the wet-sieving method described by Neubauer and Heitmann (2011). The number of MS per g soil (MS/g) was calculated from the mean of ten plates per plot.

Weather data were obtained from the German National Meteorological Service.

2.4. Data analysis

Statistical analysis was carried out using SAS 9.4. For analyzing the compatibility of biological with chemical pesticides the Glimmix procedure with random effects based on a residual likelihood was applied (*P* < 0.05). For all other experiments data showing heteroscedasticity of variance (e.g. yield) were analysed with Kruskal-Wallis nonparametric tests, comparing the means with the Wilkoxon test (*P* < 0.05).

3. Results

3.1. Laboratory experiments on the compatibility of biological with chemical pesticides

For RhizoVital® 42 fl. the number of CFUs determined directly after mixing with the pesticides as well as after 4 h incubation of the mixture, was in all cases higher than the minimum number 2.5×10^{10} spores/ml given by the producer (Fig. 1). When the CFU of the mixtures were compared immediately after mixing the CFU of the kresoxim-methyl (Discus®) mixture was significant higher than penconazole (Topas®), fosetyl-Al (Aliette®) or fenhexamid (Teldor®) mixtures. However, after 4 h incubation together with the thiacloprid insecticide Calypso® the number of CFU recovered was reduced by about 80% compared to samples plated out immediately after mixing. The two fungicides copper oxychloride (Funguran®) and kresoxim-methyl (Discus®) reduced the number of CFU's of more than 60% and 50% respectively. For all other products the reduction was less than 50%. But because of the high variance of the CFU's only quinoxyfen (Fortress®) and thiacloprid (Calypso®) were significantly different.

For Trichostar® all treatments as well as the control the germination rate was around 40% irrespective of whether they were plated immediately or after 4 h incubation. When Trichostar® was mixed with the pesticides the combination fungicide Signum®

(pyraclostrobine and boscalid) reduced the percentage of germinated spores to less than 5% which was significantly lower than all other combinations. In comparison to the control the germination was significantly higher when Trichostar® was mixed with kresoxim-methyl (Discus®). Between all other combinations no significant differences in the germination rate was obtained. Comparable results were obtained when the conidia were incubated for 4 h with the pesticide. Interestingly both kresoxim-methyl (Discus®) and trifloxystrobin (Flint®) caused a slight but not significant increase in spore germination to more than 50% (Fig. 2). The descriptive statistical values show, that after 4 h contact of biological and chemical products the variability in germination is higher than at 0 h.

3.2. Field trials

For field 1 in 2013 (Fig. 3) the yield of the RhizoVital® 42 fl. treated plants was increased by 21.3% but this could not be found in 2014. Treatment with the Mixture increased the yield in 2013 but in 2014 yield was lower than in the control plots. On the other hand Trichostar® increased the yield both years. For field 2 in 2013 (Fig. 4) Trichostar® increased the weight of the harvested strawberries by ca. 10% but in 2014 the yield was lower than the control. RhizoVital® 42 fl. treated plants produced in both years about 6% more fruits than the control. Although the yield was increased by most of the treatments, no statistical differences between treatments were obtained in both years. This was due to the high variability observed in the single treatments (Figs. 3 and 4).

A visual assessment of the plant health was made at the end of both harvesting seasons. In 2013 the number of healthy looking symptomless plants did not differ between treatments in both fields. In field 1 in 2013 about 70% of the plants were robust and healthy irrespective of the treatment. The highest number of symptomless living plants was found after application of the Mixture. However, no statistical differences between treatments was found (*P*>χ^2 0.9196/Data not shown). No visual rating was made in 2014. In field 2 in 2013 more than 95% of the plants were robust and healthy in all treatments. In 2014 the number of diseased/dead plants remained the same as in 2013 for RhizoVital® 42 fl., but doubled for control and Trichostar® however no statistical differences between treatments was found (2013 *P*>χ^2 values: symptomless 0.808; diseased 0.9742 and dead 0.8403; 2014 *P*>χ^2 values: symptomless 0.4610; diseased 0.3004 and dead 0.6292. Data not shown).

At the first sampling date (09 May 2012, before planting), the concentration of MS in field 1 was 7.2 MS/g soil. At the next sampling which was made about 14 months later clear differences

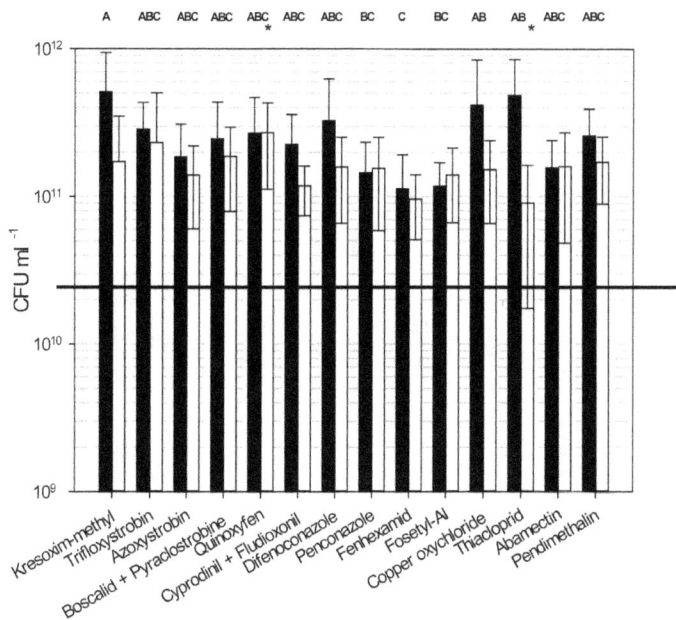

Fig. 1. Effect of chemical pesticides on survival of the microbial active ingredient of RhizoVital® FZB42 fl. Suspensions of the biological product were mixed with the pesticides and plated out immediately (black bars) or after 4 h of incubation (white bars). Means and standard deviation of N = 9 plates per treatment. Means of the log10 transformed data plated immediately after mixing with same letters are not significantly different (P < 0.05, simulated method).*: Significant differences (P < 0.05, simulated method) between the two means of the log10 transformed data. Line: calculated number of CFU per ml based on the declared product concentration.

between treatments were observed (Fig. 5). The highest concentration (41 MS/g soil) was found in the control plots and the lowest in the plots treated with the Mixture (treatment were applied by root dipping of strawberry plants before planting and soil drenches two and 10 months later). During the following months the number of MS was strongly reduced (with the number of MS for the control always higher than for the treatments) in all treatment with a small increase at the beginning of summer 2014. The reduction of MS concentration at the last sample taking with RhizoVital® 42 fl. and Trichostar® was constant even without the spring treatment. The MS concentration in field 2 before planting was 8.8 MS/g soil (sample taken on 16 August 2012). Also in this field (Fig. 6) a year after planting the number of MS increased respectively to the values before planting. In general the MS concentration values followed the trend of field 1 but with the difference that the treatments fluctuated more. At the last sampling, only in the RhizoVital® 42 fl. treatment the number of MS/g soil were below control.

3.3. Climatic aspects

Weather data are reported in Fig. 7. The temperature profile of the two sites did not differ substantially, the mean temperatures differed less than 0.5 °C. At field 1 it rained generally more than at field 2, but field 2 was irrigated when necessary. From January to March 2013 the temperatures were below 2.5 °C with several weeks below 0 °C. May 2013 was rainy with 49.3 l/m³ for field 2 and

67.5 l/m³ for field 1 in the last third of the month. In 2014 the winter was warm, with temperature means always higher than 4 °C, with no below zero daily mean.

4. Discussion

Resistance development towards chemicals and concerns over the noxious effects on human and environment safety have provided an incentive for the development of microbial control agents (Tiwari and Tripathi, 2014). After *in vitro* and *ad planta* screening of around 100 different micro-organisms the two soil strengtheners RhizoVital® 42 fl. and Trichostar® were selected to proof their influence on strawberry plants in commercial fields infested with soil-borne diseases. In addition their control capacity on *V. dahliae* MS was assessed under field conditions.

As the study was conducted in commercial fields, chemical pesticides application and other grower standard practices were followed. In an IPM strategy the tolerance of biocontrol agents (BCAs) to pesticides expands the application option of these products (Madhavi et al., 2011; Singh and Dubey, 2010). Our laboratory trials were made to obtain information whether chemical pesticides can affect the viability of the selected microbials in a practical spray schedule. Our results indicated that the concentration of viable spores of RhizoVital® FZB42 fl. was at least as high as given by the producer even after prolonged incubation with these pesticides. Archana et al. (2012) also found that the growth of *B. subtilis* was not affected by azoxystrobin and Gupta and Utkhede

145

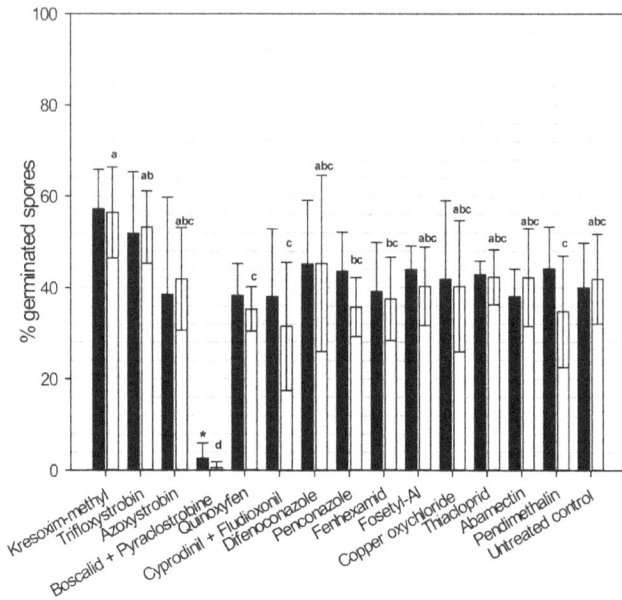

Fig. 2. Effect of chemical pesticides on survival of the microbial active ingredient of Trichostar®. Suspensions of the biological product were mixed with the pesticides and plated out immediately (black bars) or after 4 h of incubation (white bars). Means and standard deviation of N = 9 plates per treatment. *: Mean is significant different (P < 0.05, simulated method) from all other samples plated immediately after mixing. Means of the data plated after 4 h of incubation mixing with same letters are not significantly different (P < 0.05 simulated method).

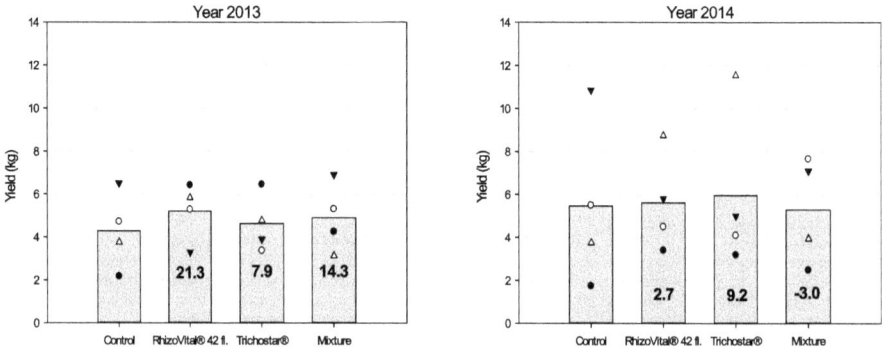

Fig. 3. Means and singular plot yield of field 1 in year 2013 (χ^2 = 0.5074, df = 3, $P>\chi^2$ = 0.9173) and 2014 (χ^2 = 0.2206, df = 3, $P>\chi^2$ = 0.9742). Columns are the means of the four singular plots (symbols). Numbers indicate the percent yield increase over the Control. This representation was chosen to show the high variability of the plots in between the treatments.

(1986) described that Fosetyl-Al did not limit multiplication and antifungal compounds production of *B. subtilis* in sterile soil. However, in other studies *B. subtilis* was not compatible with diverse fungicides (Singh and Dubey, 2010).

We confirmed the results of Harman et al. (2004) and Tripathi et al. (2013) that *Trichoderma* spp. is resistant to a variety of synthetic fungicides. In our experiments only Signum® was influencing negatively the germination rate of the fungal spores contained in Trichostar®. Signum® contains pyraclostrobin, a QoI fungicide and boscalid. Both substances influence the respiration of the fungal cell (FRAC, 2016). Other products containing QoI fungicide like chemical group oxiaminoacetate (Discus® and Flint®.

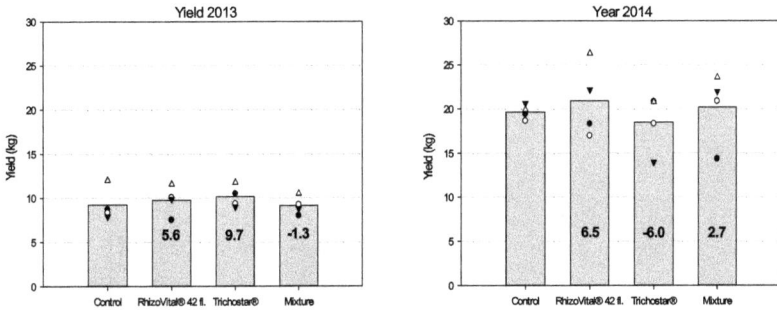

Fig. 4. Means and singular plot yield of field 2 in year 2013 ($\chi^2 = 1.6544$, df = 3, $P > \chi^2 = 0.6471$) and 2014 ($\chi^2 = 1.5221$, df = 3, $P > \chi^2 = 0.6772$). Columns are the means of the four singular plots (symbols). Numbers indicate the percent yield increase over the Control. This representation was chosen to show the variability of the plots in between the treatments.

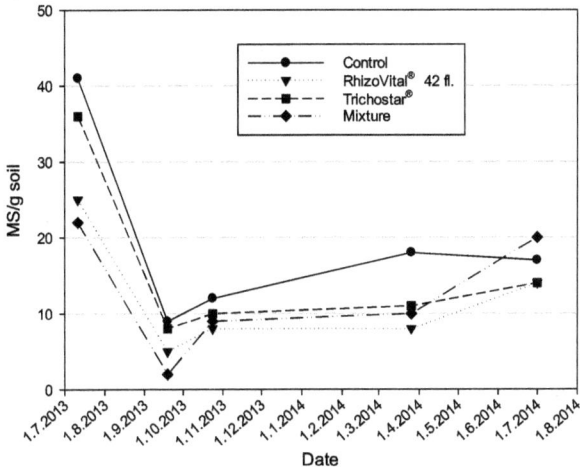

Fig. 5. Number of microsclerotia of *V. dahliae* determined in soil samples from field 1. Treatments with the microbial products were made at planting (09.05.2012; root dip) and at .07.2012 and 26.04.2013 (soil drench).

creased the percentage of germinated spores of more than 25% spective to the control, or had a neutral influence like azoxtrobin. Therefore it is likely that within Signum® boscalid influenced the germination of *T. harzianum* T58 negatively. For other tive substances various effects are described: Madhavi et al. (2011) have shown that *T. viride* mycelial growth was inhibited by e fungicides copper oxychloride and difenconazol and the herbide pendimethalin whereas Saxena et al. (2014) demonstrated that *harzianum* PBT23 was compatible with copper hydroxide and endimethalin. In our experiments with *T. harzianum*, copper xychloride slightly reduced germination and difenconazol creased it about 8% after 4 h incubation compared to the control. r pendimethalin a germination reduction of 17.2% respective to e control was seen after 4 h incubation. These different side effects can be possibly addressed to the different fungal species or rain, the concentration of the active ingredients or formulation

composition. Although this chosen test system is artificial, it is a simple method to obtain first information on the compatibility of micro-organisms with pesticides. These results have to be confirmed under field conditions.

In our field trials, the two soil strengtheners tested generally increased the yield and the number of symptomless plants respective to the untreated control.

RhizoVital® FZB42 fl. increased the yield in each field each year, once up to 21.3%. In field 2 the increase was about 6% in both years. Also the number of symptomless plants was generally higher than in the control. The positive effect of *Bacillus* species on strawberry is known. *B. subtilis* FZB24®-WG increased diverse plant parameter and fruit yield in the absence of plant pathogens but also in the presence of various diseases (Lowe et al., 2012; Tahmatsidou et al., 2006). In particular *B. amyloliquefaciens* FZB42 showed its potential in other cultivars (Chowdhury et al., 2013) and it is able to colonize

147

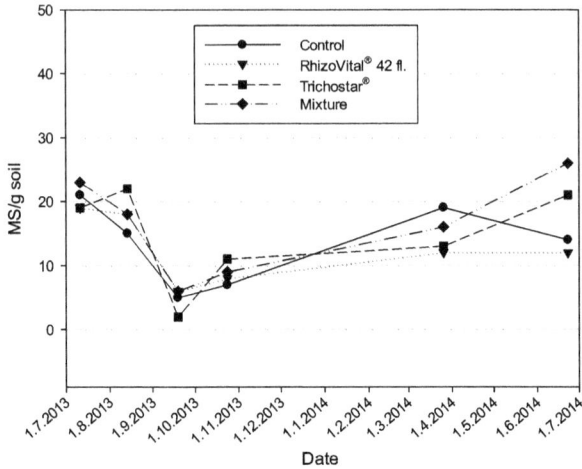

Fig. 6. Number of microsclerotia of *V. dahliae* determined in soil samples from field 2. Treatments with the microbial products were made at planting (16.08.2012; root dip) and 01.10.2012, 24.04.2013 and 16.04.2014 (soil drench).

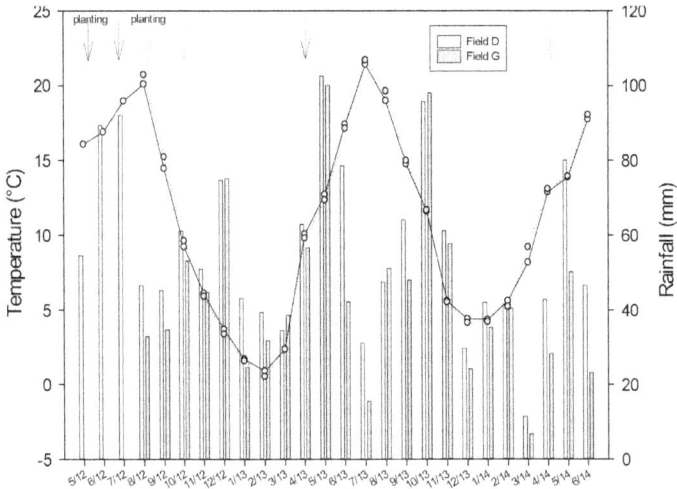

Fig. 7. Monthly mean temperature (lines with dots) and rainfall (columns) for field 1 (white) and field 2 (grey) from planting day to last picking day.

the roots of diverse plants (Fan et al., 2012). Besides, this bacterium produces secondary metabolites with antifungal and antibacterial activity and can directly inhibit pathogens growth or development (Chen et al., 2009).

In our experiments Trichostar® increased yield by 8–9% in both years for field 1, but not for field 2 in 2014, although the number of symptomless plants were lower or equal to control. *Trichoderma* exerts its activity with diverse mode of actions: mycoparasitism,

production of secondary metabolites with antimicrobial activi and various enzymes, induction of a plant defence system, ar competition for space and nutrient competition (Elad, 200 Harman et al., 2004; Howell, 2003; Singh and Singh, 2009; Tiwa and Tripathi, 2014). *Trichoderma* species are often used again *Botrytis* spp. In strawberries, disease severity was reduced by 94 by *T. harzianum* through induced resistance (Levy et al., 2015). tomato suppression of *Verticillium* wilt was due to plant grow

148

stimulation and by inhibiting penetration and migration of the pathogen through the vascular tissues of the plant (Mouria et al., 2013).

It is often reported that the application of a combination of micro-organisms can improve the plant health, in particular when these have different mode of action or biotic and abiotic requirements (Guetsky et al., 2002). However, when BCAs are applied together they can interact negatively and the efficacy can be reduced (Sylla et al., 2013). In our experiments, a simultaneous application of the two soil strengtheners none additive or synergistic effect was shown. A slight positive effect compared to the single treatments was only achieved in 2013 in the number of symptomless plants. Additional ambiguous results were obtained for the yield. Similar to our observations, researcher also found discrepant or neutral results when B. subtilis FZB24 was dual inoculated with fungi on strawberries (Lowe et al., 2012; Tahmatsidou et al., 2006).

A reduction of MS could reduce the incidence of Verticillium wilt. In our experiment in both fields high concentration of MS/g soil with seasonal variation was observed. The application of RhizoVital® FZB42 fl., Trichostar® and their mixture (except the last sampling time) could only reduce the MS/g soil of V. dahliae in one of the two fields. Mainly in laboratory or greenhouse experiments microorganisms, like Talaromyces flavus, Pseudomonas or Paenibacillus alvei strain K165 are described to reduce MS of Verticillium wilt (Antonopoulos et al., 2008; Debode et al., 2007; Fahima et al., 1992; Fravel et al., 1987; Marois et al., 1984) but reliable field data are missing.

Soil is a very dynamic environment. Adverse BCA growth conditions should be expected as a normal functioning of agriculture (Nehra and Choudhary, 2015; Trabelsi and Mhamdi, 2013). The BCA success is often unpredictable and inconstant and this is a major problem for their use (Emmert and Handelsman, 1999). Diverse trials were performed in commercial strawberry field in which yield was increased and/or symptoms of root disease were reduced by the application of BCAs or soil strengtheners. Just few papers reported positives effects in more than two consecutive years (Esitken et al., 2010; Porras et al., 2007). Although, often only one year positive effects are reported (Martin and Bull (2002) and (Kurze et al., 2001) described, that the yield enhancement from one treatment did not positively influence in the second season. When soils have similar characteristics, the efficacy discrepancies are generally explained by climatic variations (Kurze et al., 2001; Martin and Bull, 2002; Nehra and Choudhary, 2015). Thus the two sites had comparable weather conditions with differences between years. The results between fields in 2013 were not consistent, however between years in field 1 Trichostar® increased the yield of about 8 and 9% in both years and RhizoVital® FZB42 fl. about 6% in field 2. It is known that the efficacy of soil-borne BCA pathogen antagonists can be influenced by diverse abiotic enviromental factors: For T. harzianum a soil pH lower than 7 favours disease suppression (Burpee, 1990). B. amyloliquefaciens showed a better inhibition of Botrytis cinerea mycelial growth in neutral and/or alkaline pH in in vitro trials (Ahlem et al., 2012). In our trials Trichostar® showed constant results on field 1 with a pH of 6.75 and RhizoVital® FZB42 fl. in field 2 with pH 8.21. The soil texture of the two trial fields was also different. These can possibly influence the growth and performance of BCAs (Burpee, 1990).

In our study we could reach a positive effect on strawberry yield by the use of the two commercial products RhizoVital® FZB42 fl. and Trichostar®. Interestingly, the yield achieved by Trichostar® in the second year in field 1 was as high as in the first year without applying it at same time during flowering in the second year. In field 2 in which Trichostar® was applied also in the second year the yield was less than the untreated plot. For RhizoVital® FZB42 fl. the

result was reverse. With the suitable treatment schedule it would be possible to improve constantly yield and plant fitness. Anyway, growers are seeking repeated positive results, easy to handle using standard machinery and reasonably pricing (Bashan et al., 2014).

Our results based on experiments on two commercial fields in two years demonstrated the potential of biological agents. The results presented suggest, that their use has a potential to improve plant health, plant fitness and yield. Additionally, integration in an IPM strategy including synthetic pesticides seems to be possible. The mode of action is complex including direct biocontrol activity, plant growth promotion and plant strengthening. For selecting the right micro-organism it is important to consider soil characteristics and environmental factors. Additionally, their efficacy can be improved by optimizing cultural conditions like soil moisture or temperature with irrigation or by covering the surface.

Acknowledgements

Many thanks to the student assistants, to Katrin Hetebrügge, to the farmers for the allocation of the fields and the chemical pesticides and to Helmut Junge for supplying RhizoVital 42® fl. The authors thank Dr. Surendra Dara for his constructive review of the manuscript. The project was supported by funds of the Federal Ministry of Food and Agriculture (BMEL) (project number: 060E155) based on a decision of the Parliament of the Federal Republic of Germany via the Federal Office for Agriculture and Food (BLE).

References

Ahlem, H., Mohammed, E., Badoc, A., Ahmed, L., 2012. Effect of pH, temperature and water activity on the inhibition of Botrytis cinerea by Bacillus amyloliquefaciens isolates. Afr. J. Biotechnol. 11, 2210–2217.

Antonopoulos, D.F., Tjamos, S.E., Antoniou, P.P., Rafeletos, P., Tjamos, E.C., 2008. Effect of Paenibacillus alvei, strain K165, on the germination of Verticillium dahliae microsclerotia in planta. Biol. Control 46, 166–170.

Archana, S., Manjunath, H., Nijagunaiah, R., Raguchander, T., 2012. Compatibility of azoxystrobin 23 SC with biocontrol agents and insecticides. Madras Agric. J. 99, 374–377.

Bashan, Y., de-Bashan, L.E., Prabhu, S.R., Hernandez, J.-P., 2014. Advances in plant growth-promoting bacterial inoculant technology: formulations and practical perspectives (1998–2013). Plant Soil 378, 1–33.

Berg, G., 2007. Biological Control of Fungal Soilborne Pathogens in Strawberries.

Bundesamt, S., 2016. Land- und Forstwirtschaft, Fischerei Gemüseerhebung - Anbau und Ernte von Gemüse und Erdbeeren - 2015, vol. 2016. Statistisches Bundesamt. Wiesbaden, pp. 93.

Burpee, L.L., 1990. The influence of abiotic factors on biological control of soilborne plant pathogenic fungi. Can. J. Plant Pathology-Revue Can. De Phytopathologie 12, 308–317.

Chen, X.-H., Koumoutsi, A., Scholz, R., Borriss, R., 2009. More than anticipated - production of antibiotics and other secondary metabolites by Bacillus amyloliquefaciens FZB42. J. Mol. Microbiol. Biotechnol. 16, 14–24.

Chowdhury, S.P., Dietel, K., Raendler, M., Schmid, M., Junge, H., Borriss, R., Hartmann, A., Grosch, R., 2013. Effects of Bacillus amyloliquefaciens FZB42 on lettuce growth and health under pathogen pressure and its impact on the rhizosphere bacterial community. Plos One 8.

Colla, P., Gilardi, G., Gullino, M.L., 2012. A review and critical analysis of the European situation of soilborne disease management in the vegetable sector. Phytoparasitica 40, 515–523.

Conti, S., Villari, G., Faugno, S., Melchionna, G., Somma, S., Caruso, G., 2014. Effects of organic vs. conventional farming system on yield and quality of strawberry grown as an annual or biennial crop in southern Italy. Sci. Hortic. 180, 63–71.

Debode, J., De Maeyer, K., Perneel, M., Pannecoucque, J., De Backer, G., Höfte, M., 2007. Biosurfactants are involved in the biological control of Verticillium microsclerotia by Pseudomonas spp. J. Appl. Microbiol. 103, 1184–1196.

Dubey, S.C., Bhavani, R., Singh, B., 2011. Integration of soil application and seed treatment formulations of Trichoderma species for management of wet root rot of mungbean caused by Rhizoctonia solani. Pest Manag. Sci. 67, 1163–1168.

Elad, Y., 2000. Biological control of foliar pathogens by means of Trichoderma harzianum and potential modes of action. Crop Prot. 19, 709–714.

Elad, Y., Chet, I., Henis, Y., 1981. Biol. control Rhizoct. solani Strawb. fields by Trichoderma harzianum Plant Soil 60, 245–254.

Emmert, E.A.B., Handelsman, J., 1999. Biocontrol of plant disease: a (Gram-) positive perspective. Fems Microbiol. Lett. 171, 1–9.

Esitken, A., Yildiz, H.E., Ercisli, S., Donmez, M.F., Turan, M., Gunes, A., 2010. Effects of plant growth promoting bacteria (PGPB) on yield, growth and nutrient contents of organically grown strawberry. Sci. Hortic. 124, 62–66.

Fahima, T., Madi, L., Henis, Y., 1992. Ultrastructure and germinability of Verticillium dahliae microsclerotia parasitized by Talaromyces flavus on agar medium and in treated soil. Biocontrol Sci. Technol. 2, 69–78.

Fan, B., Borriss, R., Bleiss, W., Wu, X., 2012. Gram-positive rhizobacterium Bacillus amyloliquefaciens FZB42 colonizes three types of plants in different patterns. J. Microbiol. 50, 38–44.

FAO, 2016. FAOSTAT Database. http://faostat3.fao.org/download/Q/QC/E.

FRAC, 2016. FRAC Code List ©*2016: Fungicides Sorted by Mode of Action (Including FRAC Code Numbering). http://www.frac.info/publications/downloads.

Fravel, D.R., Kim, K.K. Papavizas, G.C., 1987. Viability of microsclerotia of Verticillium dahliae reduced by metabolite produced by Talaromyces flavus. Phytopathology 77, 616–619.

Freeman, S., Minz, D., Kolesnik, I., Barbul, O., Zveibil, A., Maymon, M., Nitzani, Y., Kirshner, B., Rav-David, D., Bibi, A., Dag, A., Shafir, S., Elad, Y., 2004. Trichoderma biocontrol of Colletotrichum acutatum and Botrytis cinerea and survival in strawberry. Eur. J. Plant Pathol. 110, 361–370.

Guetsky, R., Shtienberg, D., Elad, Y., Fischer, E., Dinoor, A., 2002. Improving biological control by combining biocontrol agents each with several mechanisms of disease suppression. Phytopathology 92, 976–985.

Gupta, V.K., Utkhede, R.S., 1986. Factors affecting the production of antifungal compounds by Enterobacter aerogenes and Bacillus subtilis, antagonists of Phytophthora cactorum. J. Phytopathology-Phytopathologische Zeitschrift 117, 9–16.

Harman, G.E., Howell, C.R., Viterbo, A., Chet, I., Lorito, M., 2004. Trichoderma species - opportunistic, avirulent plant symbionts. Nat. Rev. Microbiol. 2, 43–56.

Harris, D.C., Yang, J.R., 1996. The relationship between the amount of Verticillium dahliae in soil and the incidence of strawberry wilt as a basis for disease risk prediction. Plant Pathol. 45, 106–114.

Howell, C.R., 2003. Mechanisms employed by Trichoderma species in the biological control of plant diseases: the history and evolution of current concepts. Plant Dis. 87, 4–10.

Kredics, L., Antal, Z., Manczinger, L., Szekeres, A., Kevei, F., Nagy, E., 2003. Influence of environmental parameters on Trichoderma strains with biocontrol potential. Food Technol. Biotechnol. 41, 37–42.

Kurze, S., Bahl, H., Dahl, R., Berg, G., 2001. Biological control of fungal strawberry diseases by Serratia plymuthica HRO-C48. Plant Dis. 85, 529–534.

Levy, N.O., Harel, Y.M., Haile, Z.M., Elad, Y., Rav-David, E., Jurkevitch, E., Katan, J., 2015. Induced resistance to foliar diseases by soil solarization and Trichoderma harzianum. Plant Pathol. 64, 365–374.

Lowe, A., Rafferty-McArdle, S.M., Cassells, A.C., 2012. Effects of AMF- and PGPR-root inoculation and a foliar chitosan spray in single and combined treatments on powdery mildew disease in strawberry. Agric. Food Sci. 21, 28–38.

Maas, J.L., 1998. Compendium of Strawberry Diseases. APS Press, St. Paul, MN.

Madhavi, G.B., Bhattiprolu, S.L., Reddy, V.B., 2011. Compatibility of biocontrol agent Trichoderma viride with various pesticides. J. Hortic. Sci. 6, 71–73.

Marois, J.J., Fravel, D.R., Papavizas, G.C., 1984. Ability of Talaromyces flavus to occupy the rhizosphere and its interaction with Verticillium dahliae. Soil Biol. Biochem. 16, 387–390.

Martin, F.N., Bull, C.T., 2002. Biological approaches for control of root pathogens of strawberry. Phytopathology 92, 1356–1362.

Mouria, B., Ouazzani Touhami, A., Douira, A., 2013. Effet du compost et de Trichoderma harzianum sur la suppression de la verticilliose de la tomate. J. Appl. Biosci. 70, 5531–5543.

Nam, M.H., Kim, H.S., Lee, H.D., Whang, K.S., Kim, H.G., 2014. Biological control of anthracnose crown rot in strawberry using Bacillus velezensis NSB-1. In: Zhang, Y.T., Maas, J. (Eds.), Acta Horticulturae, pp. 685–688.

Nehra, V., Choudhary, M., 2015. A review on plant growth promoting rhizobacteria acting as bioinoculants and their biological approach towards the production of sustainable agriculture. J. Appl. Nat. Sci. 7, 540–556.

Neubauer, C., Heitmann, B., 2011. Quantitative detection of Verticillium dahliae in soil as a basis for selection of planting sites in horticulture. J. für Kulturpflanzen 63, 1–8.

Porras, M., Barrau, C., Arroyo, F.T., Santos, B., Blanco, C., Romero, F., 2007. Reduction of Phytophthora cactorum in strawberry fields by Trichoderma spp. and soil solarization. Plant Dis. 91, 142–146.

Saxena, D., Tewari, A.K., Dinesh, R., 2014. The in vitro effect of some commonly used fungicides, insecticides and herbicides for their compatibility with Trichoderma harzianum PBT23. World Appl. Sci. J. 31, 444–448.

Singh, B., Dubey, S.C., 2010. Bioagent based integrated management of Phytophthora blight of pigeonpea. Archives Phytopathology Plant Prot. 43, 922–929.

Singh, H.B., Singh, D.P., 2009. From biological control to bioactive metabolites: prospects with Trichoderma for safe human food. Pertanika J. Trop. Agric. Sci. 32, 99–110.

Sylla, J., Alsanius, B.W., Krueger, E., Becker, D., Wohanka, W., 2013. In vitro compatibility of microbial agents for simultaneous application to control strawberry powdery mildew (Podosphaera aphanis). Crop Prot. 51, 40–47.

Tahmatsidou, V., O'Sullivan, J., Cassells, A.C., Voviatzis, D., Paroussi, G., 2006. Comparison of AMF and PGPR inoculants for the suppression of Verticillium wilt of strawberry (Fragaria x ananassa cv. Selva). Appl. Soil Ecol. 32, 316–324.

Tiwari, Tripathi, 2014. Kapitel 9. The Multifaceted Role of the Trichoderma System in Biocontrol: Biological Controls for Preventing Food Deterioration: Strategies for Pre- and Postharvest Management.

Trabelsi, D., Mhamdi, R., 2013. Microbial inoculants and their impact on soil microbial communities: a review. Biomed Res. Int. http://dx.doi.org/10.1155/2013/863240. **Article Number: 863240.**

Tripathi, P., Singh, P.C., Mishra, A., Chauhan, P.S., Dwivedi, S., Bais, R.T., Tripathi, R.D., 2013. Trichoderma: a potential bioremediator for environmental clean up. Clean Technol. Environ. Policy 15, 541–550.

Vestberg, M., Kukkonen, S., Kim, H., Saari, K., Hurme, T., 2008. Effect of cropping system and peat amendment on strawberry growth and yield. Agric. Food Sci. 17, 88–101.

Selected micro-organisms in greenhouse and controlled field trials against *Verticillium dahliae* and *Phytophthora cactorum* on strawberries (2017).

Bisutti, I.L.

Extract of the final project report FKZ 2811NA012

Isabella L. Bisutti and Dietrich Stephan (2014): Einsatz mikrobiologischer Präparate zur Regulierung der bodenbürtigen Erdbeerkrankheiten *Verticillium dahliae* und *Phytophthora cactorum* sowie des Erdbeerblütenstechers

Selected micro-organisms in greenhouse and controlled field trials against *Verticillium dahliae* and *Phytophthora cactorum* on strawberries

Bisutti Isabella L.

Material and methods

Microbial material for greenhouse and field trials

Metarhizium brunneum Ma43, provided by JKI-Institute for Biological Control, was produced at the institute facilities. It was grown in a solid fermenter (Prophyta) on rice:oat (1:5) substrate for 14 days at 25 °C with air supply. After harvest, it was kept at 4-7°C until use. The spores were gained by agitating the solid material in 0.1% Tween[®] 80 solution. The germination capacity was regularly tested and the concentration used in the trials was of 10^5 germinating spores/ml. The proper concentration was adjusted with tap water. RhizoVital[®] 42 fl. and Trichostar[®] were prepared as described in supplement 4, section chapter 2.1. The combination of all three (here after referred to as Mixture) was prepared as described in supplement 4, section chapter 2.1 with the addition of *M. brunneum* Ma43.

The pathogen *P. cactorum* A1, provided by JKI-Institute for Plant Protection in Horticulture and Forests, was routinely grown on V8 agar (V8 broth and 15 g agar Merck, Germany). The mycelia was then scraped from the surface and added to V8 broth (300 ml vegetable juice (Viva Vital Netto, Germany), 4.5 g $CaCO_3$ (Merck, Germany), 30 mg β-Sitosterol (Fluka, Germany) and water to one litre) and kept at 20 °C for 2-4 weeks. This suspension was then used to inoculate an autoclaved vermiculite wheat bran mix (100.0 g vermiculite, 100.0 g wheat bran, 50 ml vegetable juice cleared with $CaCO_3$ (Merck, Germany)). After 6 weeks in the dark at 20 °C the mixture was stored at 4 °C until used. The pathogen *V. dahliae*, was provided by JKI-Institute for Breeding Research on Fruit Crops, was a mix of different strains with various levels of virulence. Routinely they were grown on CD agar (35.0 g Czapek-Dox broth (Otto Nordwald, Germany) and 15.0 g agar (Roth)). The inoculum was prepared by adding pieces of grown agar to 100 ml CD broth and fermented on a rotary shaker at 20 °C by 100 rpm in the dark. After 14 days 20 ml of the suspension was added to an autoclaved mix of sand and rye flour (210.0 g sand, 15.0 g rye flour in health food quality and 20 ml de-ionised water). After 6 weeks at 20 °C in the dark the mix was air dried and sieved (1 mm) and kept at 4 °C until use (Heitmann, personal communication).

Laboratory experiments on the compatibility of Ma43 with chemical pesticides

For the methodology see supplement 4, section chapter 2.2. *M. brunneum* Ma43 was tested at concentration of 10^7 spores ml in 0.1% Tween[®] 80 so that centrifugation after contact was not necessary.

Greenhouse trials 2013

Commercial strawberry plants (cv. "Honeoye", Kraege Beerenpflanzen GmbH & Co. KG, Germany) were delivered as cold stored bare rooted and kept at 4 °C until use. Before planting, the roots were cut to a length of 13 cm and then dipped for 15 min in 2 l of the antagonist suspension or in 2 l tap water used as control. There were an untreated control and four treatments: RhizoVital® 42 fl., Trichostar®, *M. brunneum* Ma43 and the Mixture. Afterwards, the strawberries were planted in 30 cm diameter plastic pots filled with 0.7 kg of a non-sterile zero soil (Fruhstorfer Erde, Germany):perlite (3:1) mixture. For the trials with pathogen inoculation, 1.5% of *V. dahliae* or 5% of *P. cactorum* inoculum was added to the soil and mixed for 15 min at maximum speed (level 8) with a dough blender. Six plants per antagonist and disease were planted and the pots were put in a Euro tray. After watering, the plants were placed in a greenhouse where temperature was monitored. The experiments started in spring 2013 and the plants were maintained under natural light conditions.

In one-week intervals, three trials were set and the Euro trays were regularly rotated during the trial time. When 80% of the plants were at BBCH 61, about four weeks after planting, 200 ml of the antagonistic suspension was added pro plant. Plants were watered when needed and fertilized four times (6, 10, 14 and 18 weeks after planting) with liquid Naturbell Pflanzendünger (COMPO, Germany). Dead plants were removed and substituted with plants of the same age to maintain the same microclimate conditions for all repetitions. *Chrysoperla carnea* s.l larvae and subsequently Spruzit® (4.59 g/l pyrethrin and 825.3 g/l rape oil) were used to reduce aphid invasion. After six months the plants were destructively sampled and the soil from the roots was washed off using tap water and the excess water was left to dry. The following parameters were determined: plant fresh and dry weight, new roots and aerial part dry weight, length of the longest root, number of green leaves, number and dry weight of daughter plants. To determine the dry weight, the aerial parts were dried at 60 °C for 90 h and the roots and shoot for 115 h. Biomass increase (BI) was calculated as:

$$BI = \frac{fresh\ weight\ at\ the\ end\ of\ the\ trial - initial\ weight\ of\ the\ bare\ root\ plant}{initial\ weight\ of\ the\ bare\ root\ plant} \times 100$$

Each combination of antagonists, pathogen and controls was replicated one time in a randomized design.

Pathogen inoculated field trials (2012-2014)

The trial was conducted at the JKI-Institute for Biological Control in Darmstadt (Hesse). In spring 2012 compost was added to the field (sand 92.9%, silt and clay 7.1%; pH 8.37; with artificial irrigation; previous crop: strawberry). Commercial strawberry plants (cv. "Honeoye", Kraege Beerenpflanzen GmbH & Co. KG, Germany) were delivered as frigo (cold stored bare rooted) and kept at 4 °C till use. Each plot contained three rows with 25 strawberry plants each. The middle row had no disease inoculation, the external ones were artificially inoculated by adding ca. 2 g of *V. dahliae* or 4 g of *P. cactorum* to the planting hole. The distance between plants was 25 cm and between rows 50 cm. The distance between treatments was 90 cm and covered with plastics to avoid weeds. The plants were prepared as described for the greenhouse trials. Two months after planting and in the spring of 2013, a treatment with the antagonists was performed. For that treatment, 5 l of the suspension was used to water one row. After a harsh winter, in the spring 2013, it was decided to substitute the dead plants with new ones treated as described above, and planted in the same hole as previously occupied by the dead ones.

Fruit yield was assessed by picking all ripe and healthy fruits. In 2013 the harvest started at the beginning of June, which was abnormally late. In 2014 the harvesting time started two weeks before 2013. The field was picked 4 times in 2013 and 10 times in 2014 by student assistants. In 2013 and 2014 the field was visually rated three times in 2013 and at the end of the harvesting season in 2014. The rating scale is shown in the picture below, with rating 6 for dead plants.

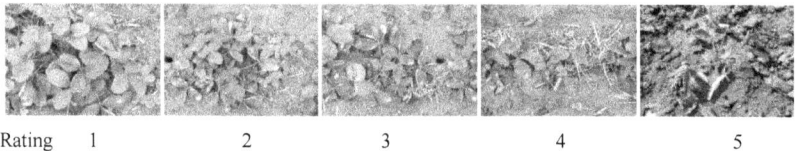

Rating   1   2   3   4   5

Harvest and rating in 2013 were made separate for plants planted in 2012 and planted in 2013. In 2014 data was pooled. For each treatment (25 plants each) four repetitions in a random block were assessed. Weather data were recorded at the JKI-Institute's weather station and soil analysis were made at the JKI-Institute's laboratory.

Data analysis

Statistical analysis was carried out using SAS 9.4. For analyzing the compatibility of biological with chemical pesticides the Glimmix procedure with random effects based on a residual likelihood was applied ($P < 0.05$). For all other experiments data showing heteroscedasticity of variance (*e.g.* yield) were analysed with Kruskal-Wallis nonparametric tests, comparing the means with the Wilkoxon test ($P < 0.05$).

Results

Laboratory experiments on the compatibility of Ma43 with chemical pesticides

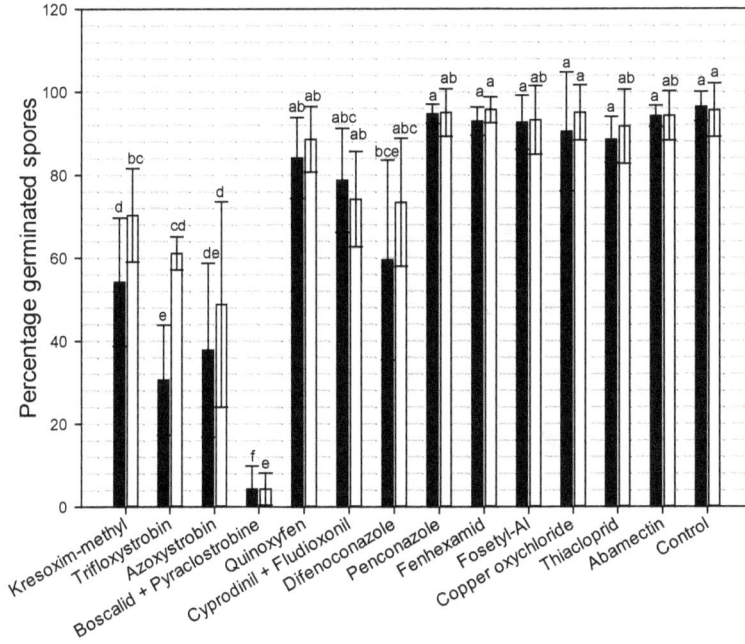

Figure A1: Effect of chemical pesticides on survival of *Metarhizium brunneum* Ma43. Suspensions of the entomopathogenic fungi were mixed with the pesticides and plated out immediately (black bars) or after 4 h of incubation (white bars). Means and standard deviation of N= 9 plates per treatment. Means of the data with the same letters are not significantly different (P < 0.05).

Greenhouse trials 2013

Total plants: 270 = 5 treatments x 3 pathogens x 3 repetitions x 6 plants
Three replication with one-week interval (n=6 plants x 3 repetitions), date: first planting (FP)   18.04; second planting (SP) 25.04 and third planting (TP) 02.05.
Additional treatments dates: FP 17.05; SP 24.05 and TP 31.05.
Fruits picked when ripe.
Daughter plants were harvested: FP 25.09; SP 02.10 and TP 09.10.
Mother plants were harvested and the growth parameters were assessed by destructive yield: FP 16.10; SP 23.10 and TP 30.10.

Table A1: Growth parameter recorded at the end of the greenhouse trial.

	BI (%)	Plant dry weight (g/plant)	Leaf dry weight (g/plant)	Number of green leaves (plant)&	Root dry weight (g/plant)§	Root length (plant)$
C	332 (± 132)b	11.08 (± 2.48)	6.84 (± 1.40)	15.2 (± 3.6)	1.18 (± 0.60)b	27.8 (± 6.2)a
R	382 (± 126)ab	11.29 (± 2.23)	6.98 (± 1.19)	15.0 (± 3.2)	1.23 (± 0.40)b	28.3 (± 3.9)ab
T	431 (± 132)a	12.33 (± 2.40)	7.13 (± 1.27)	15.4 (± 3.9)	1.80 (± 0.72)a	31.3 (± 5.8)b
M	352 (± 172)ab	12.28 (± 2.89)	7.02 (± 1.58)	14.6 (± 4.9)	1.83 (± 0.71)a	30.6 (± 6.2)ab
X	350 (± 128)ab	11.64 (± 2.14)	6.72 (± 1.09)	14.6 (± 3.1)	1.48 (± 0.66)ab	29.9 (± 5.8)ab
Df	4	4	4	4	4	4
χ^2	4.6625	2.9210	1.0154	0.39827	15.5335	6.9424
Pr>χ^2	0.3237	0.5711	0.9075	0.9839	0.0037	0.1390
Vd	320 (± 131)	11.73 (± 2.47)ab	6.96 (± 1.85)	14.1 (± 4.2)ab	1.45 (± 0.54)ab	26.3 (± 5.4)
Vd x R	377 (± 109)	11.14 (± 2.96)b	6.48 (± 1.36)	15.1 (± 5.7)b	1.39 (± 0.58)b	26.7 (± 5.2)
Vd x T	403 (± 197)	11.71 (± 3.48)ab	6.88 (± 1.65)	14.6 (± 4.1)ab	1.26 (± 0.66)b	25.8 (± 5.7)
Vd x M	308 (± 85)	12.56 (± 3.16)ab	7.54 (± 2.42)	14.8 (± 3.2)ab	1.22 (± 0.64)b	24.0 (± 5.8)
Vd x X	346 (± 94)	13.11 (± 2.19)a	7.13 (± 0.98)	16.7 (± 2.3)a	1.73 (± 0.42)a	24.7 (± 6.2)
Df	4	4	4	4	4	4
χ^2	2.7493	6.0017	3.4018	6.0540	9.9998	2.1321
Pr>χ^2	0.6006	0.1990	0.4930	0.1952	0.0404	0.7115
Pc	127 (± 161)b*	11.30 (± 2.77)ab	7.43 (± 1.74)a	7.5 (± 8.2)*	0.37 (± 0.37)a*	15.6 (± 3.1)c*
Pc x R	241 (± 163)ab	10.48 (± 1.55)a	6.43 (± 0.91)b	10.6 (± 6.1)	0.52 (± 0.44)ab	17.3 (± 3.5)bc
Pc x T	247 (± 120)ab	8.98 (± 1.73)c	5.70 (± 1.11)a	9.3 (± 4.7)	0.39 (± 0.26)a	17.9 (± 2.8)b
Pc x M	235 (± 112)a	10.89 (± 2.09)ab	7.11 (± 1.15)a	11.8 (± 4.1)	0.66 (± 0.38)b	21.9 (± 4.3)a
Pc x X	302 (± 78)a	9.86 (± 2.28)bc	6.54 (± 1.55)ab	12.1 (± 5.7)	0.71 (± 0.54)b	18.7 (± 4.3)b
Df	4	4	4	4	4	4
χ^2	8.8811	13.2422	14.3414	5.5574	7.7605	21.6138
Pr>χ^2	0.0641	0.0102	0.0063	0.2347	0.1008	0.0002

C=no pathogen inoculation; R=RhizoVital® 42 fl.; T=Trichostar®; M=*M. brunneum* Ma43; X=Mixture (R+T+M); Vd=*V. dahliae* and Pc=*P. cactorum*

BI%=Percent biomass increase, calculated with the fresh weight of the whole plant and the produced stolons; n=6 plants*2 repetition (second and third trial)

For no and *V dahliae* inoculation all plants were alive at the end of the trial. For *P. cactorum*: only Pc 8; RhizoVital® 42 fl. 4; Trichostar® 2; *M. brunneum* Ma43 0 and Mixture 2 plants were dead

* Means of controls (no antagonistic treatment) is significant different (P < 0.05).

& Number of green leaves at harvest. For death plants the value used was zero because no green leaves were left

§ Mean of the weight of newly produced roots. For death plants the value used was zero because no fresh roots were produced

$ Mean of the measure of the longest root in cm. For death plants the value used was 13 cm, the value at planting

156

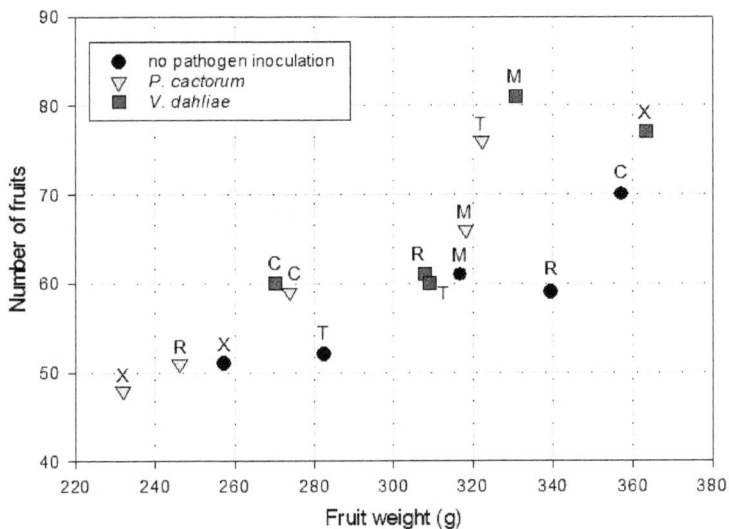

Figure A2: Distribution of fruit weight and quantity in relation to pathogen and treatment applied.

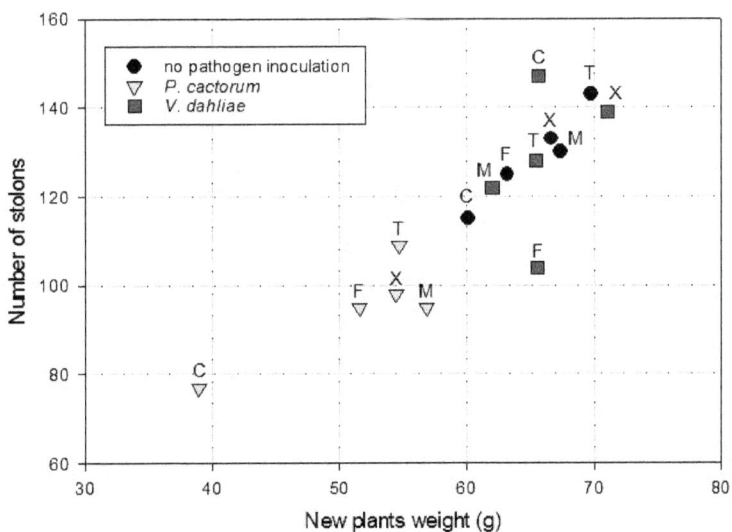

Figure A3: Distribution of the dry weight of new daughter plants and stolon quantity in relation to pathogen and treatment applied.

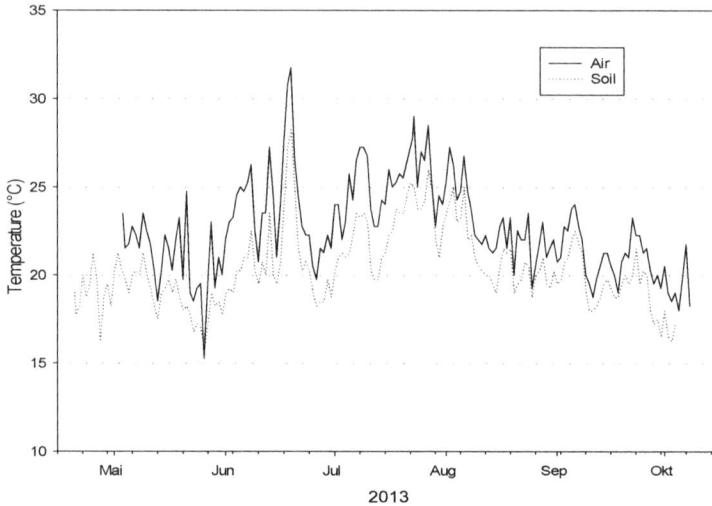

Figure A4: Temperature profile from air and soil during greenhouse trials from April to October 2013. Data are the mean of the day.

Pathogen inoculated field trials (2012-2014)

Total plants: 1500 = 5 treatments x 3 pathogens x 4 replicates x 25 plants.
Plants were planted on 23.08.2012. Substitution of dead plants on 25.05.2013.
Additional treatments date: 01.11.2012 and 06.06.2013.
Fruits picked when ripe.

Table A2: Yield of the field in controlled conditions.

	2013 (g/plant)§						2014 (kg)		
	P. cactorum		C		*V. dahliae*		*P. cactorum*	C	*V. dahliae*
	old	new	old	new	old	new			
Control	18.9	26.4	29.1	23.8	24.2	24.6	26.2	**32.4**	**33.6**
RhizoVital® 42 fl.	15.0	29.5	23.2	30.6	14.1	25.2	24.0	28.1	28.7
Trichostar®	18.7	25.5	29.8	24.0	19.2	24.6	24.7	29.6	28.3
M. brunneum **Ma43**	10.0	27.5	22.5	26.0	22.3	26.6	24.1	28.0	27.9
Mixture	14.5	29.2	26.8	28.3	22.1	27.0	**26.7**	29.8	31.1

C: no pathogen inoculation; Control: no antagonistic treatment
2013: The reported data are separated in old and new because the number of plants differed between treatments and rows; § weight of picked strawberry per plant. No statistical analysis because the number of plants were too different.
2014: Total yield of the treatments (sum of 4 plots with 25 plants) χ^2=13.2066, df=4, P> χ^2 = 0.5103

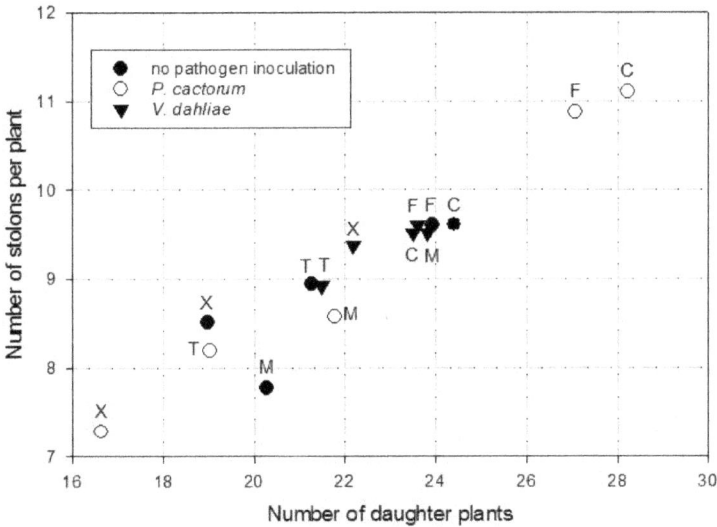

Figure A5: Distribution of the dry weight of new daughter plants and stolon quantity in relation to pathogen and treatment applied. The values were calculated in relation to the living plants in 2013. Harvested between 24.07. and 07.08.

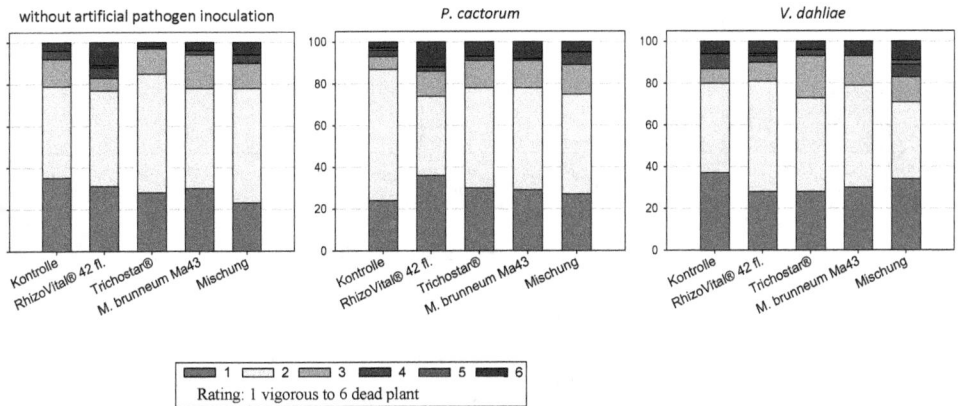

Figure A6: Influence of the application of antagonistic micro-organism on plant vigor/vitality (Rating scale 1 to 6) in June 2014 (Kontrolle=no treatment applied; Mischung=Mixture of the two bioproducts with *M. brunneum* Ma43 spores).

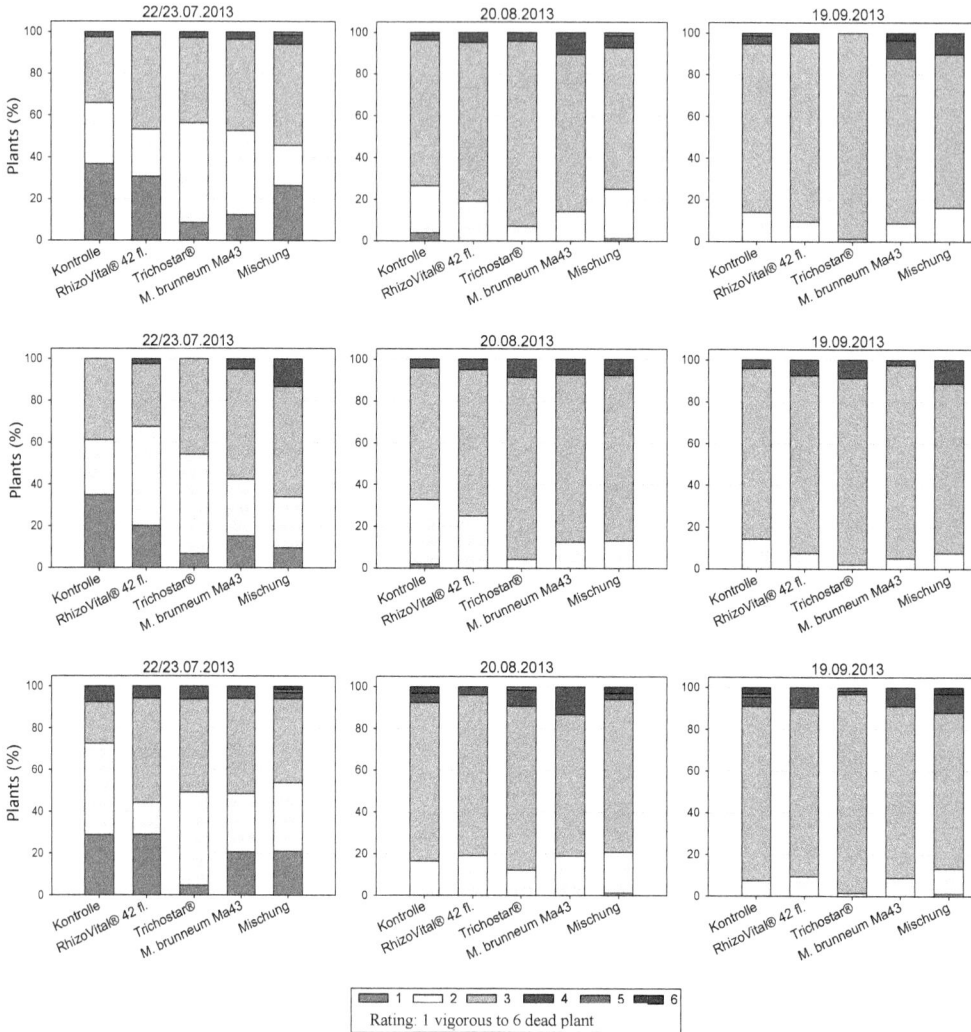

Figure A7: Influence of the application of antagonistic microorganism on plant vigour/vitality (Rating scale 1 to 6) of the strawberry planted in 2012 at different dates (upper row without artificial pathogen inoculation, middle row inoculated with *P. cactorum* and under row inoculated with *V. dahliae*) (Kontrolle=no treatment applied; Mischung=Mixture of the two bioproducts with *M. brunneum* Ma43 spores).

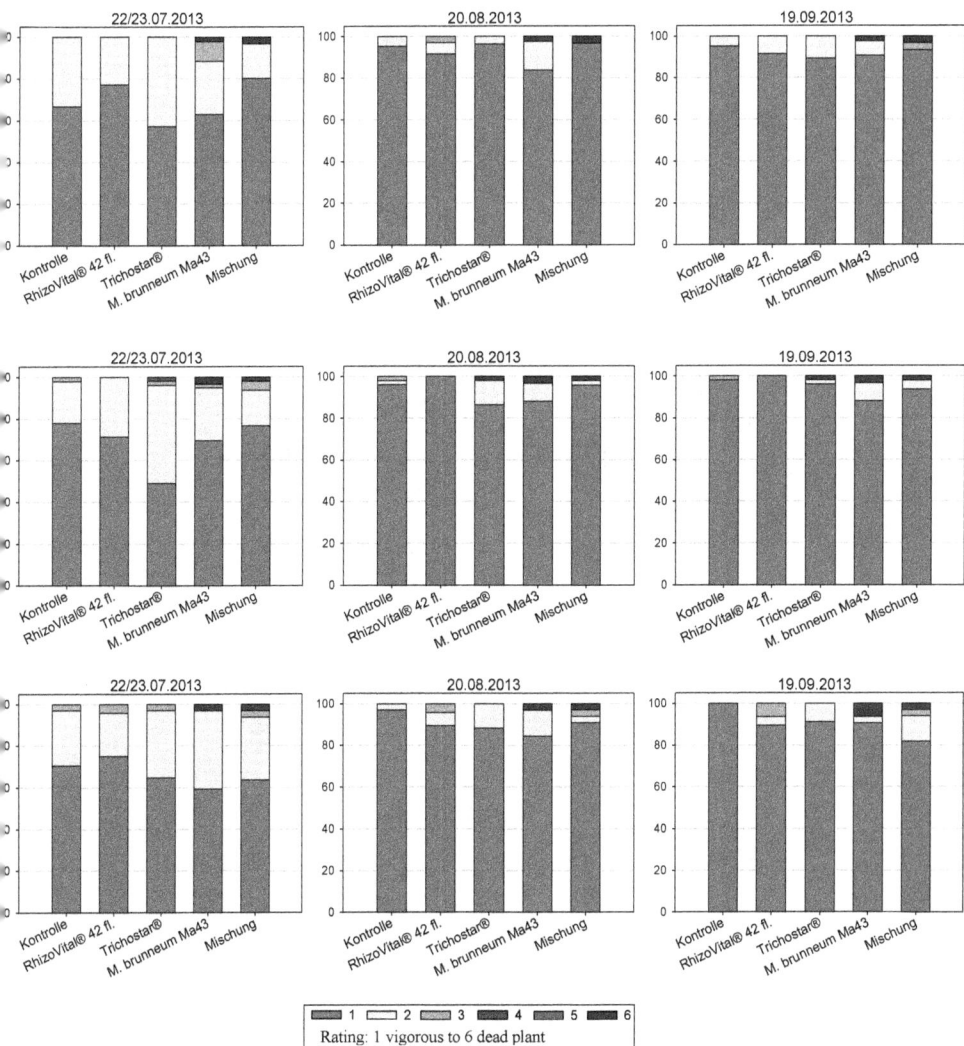

Rating: 1 vigorous to 6 dead plant

Figure A8: Influence of the application of antagonistic microorganism on plant vigour/vitality (Rating scale 1 to 6) of the strawberry planted in 2013 at different dates (upper row without artificial pathogen inoculation, middle row inoculated with *P. cactorum* and under row inoculated with *V. dahliae*) (Kontrolle=no treatment applied; Mischung=Mixture of the two bioproducts with *M. brunneum* Ma43 spores).

161

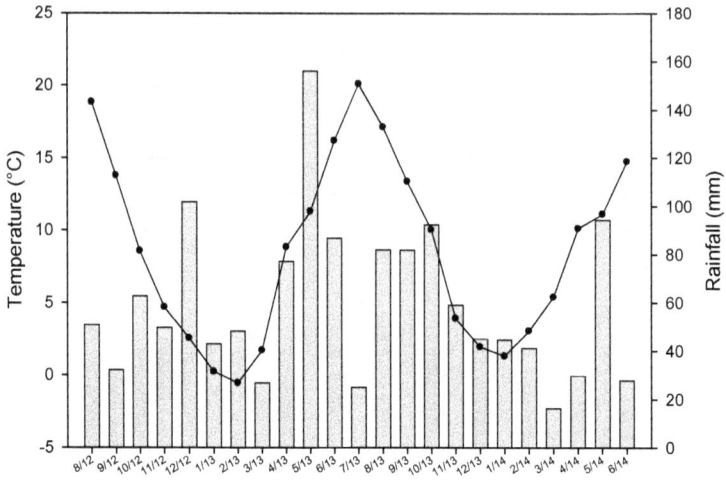

Figure A9: Monthly rainfall (columns) and mean temperature (line with dots) during the field trial from August 2012 to June 2014.

ACKNOWLEDGEMENTS

I would like to kindly thank my supervisor Prof. Dr. Carmen Büttner for accepting me as doctoral candidate, providing me support and encouragement and the opportunity to meet the division of Phytomedicine, which was always friendly and helpful when I travelled to Berlin.

I would like to acknowledge my advisor Dr. Dietrich Stephan for his scientific and personal support, the interesting scientific discussions and the opportunities to attend to different conferences. I want to acknowledge Juliana Pelz for her remarkable laboratory and field work, for organizing the kindergarten of all the nice and helpful student assistants. They were so many and I want to thank them all too for the work they made and the funny times together.

Many thanks to all the personnel of the Biological Institute in Darmstadt for their support and the nice time. Many thanks also to Irmgard, which welcomed me with open hands.

Thanks to SafeCrop and Böln for financial support and the JKI for the possibility to work in their facilities.

I am extremely grateful to my family, especially my mother, for the incredible mental support in all these years, the believe in my capabilities and the patience and encouragement during my hard times.

I would like to express my gratitude to all the persons that are not mentioned by name but helped me on my scientific and personal way.

In der Reihe *Berliner ökophysiologische und phytomedizinische Schriften* sind bisher erschienen:

Band 01:   Mohammad Mahir Uddin (2009)
            Chemical ecology of mustard leaf beetle Phaedon cochleariae (F.).
            ISBN 978-3-89959-848-3.

Band 02:   Ilir Morina (2009)
            Entwicklung von Verfahren zur Rekultivierung der Aschedeponie des
            Braunkohlekraftwerks in Prishtina (Kosovo).
            ISBN 978-3-89959-872-8.

Band 03:   Melanie Wiesner (2009)
            Veränderungen gesundheitsrelevanter Inhaltsstoffe in *Parthenium
            hysterophorus* L. in Abhängigkeit von der Pflanzengröße und Klimafaktoren.
            ISBN 978-3-89959-880-3.

Band 04:   Fransika Rohr (2009)
            Variabilität aliphatischer Glucosinolate in *Arabidopsis thaliana*-Ökotypen und
            deren Einfluss auf die Wirtspflanzeneignung von zwei folivoren Insektenarten.
            ISBN 978-3-89959-884-9.

Band 05:   Jutta Buchhop (2009)
            Characterization of phylogenetically diverse CLRV-isolates by RFLP and
            research into identification of two isometric viruses.
            ISBN 978-3-89959-929-9.

Band 06:   Nora Koim (2010)
            Urban sprawl, land cover change and forest fragmentation – Case study
            Pereira, Colombia.
            ISBN 978-3-89959-955-8.

Band 07:   Nadja Förster (2010)
            Eignung unterschiedlicher salicylathaltiger *Salix*-Klone für die
            Arzneimittelindustrie.
            ISBN 978-3-89959-964-0.

Band 08:   Jana Gentkow (2010)
            Cherry leaf roll virus (CLRV): Charakterisierung ausgewählter Virusisolate
            unter besonderer Berücksichtigung des viralen Hüllproteins.
            ISBN 978-3-89959-976-3.

Band 09:   Ahmad Fakhro (2010)
            Interaction of Pepino mosaic virus (PepMV) and fungal root endophytes with
            tomato hosts (*Lycopersicum esculentum* Mill.).
            ISBN 978-3-89959-995-4.

Band 10:   Stefan Irrgang (2010)
            Mikro- und makroskopische Untersuchungen an Veredelungsstellen von
            Straßenbäumen im Hinblick auf die Beeinflussung ihrer Bruchsicherheit.
            ISBN 978-3-89959-998-5.

Band 11:   Julia Jahnke (2010)
            Guerilla Gardening anhand von Beispielen in New York, London und Berlin.
            ISBN 978-3-86247-001-3.

Band 12:   Astrid Karoline Günther (2010)
Analysen zur Intensität der Pflanzenschutzmittel-Anwendung und Aufklärung ihrer Einflussfaktoren in ausgewählten Ackerbaubetrieben.
ISBN 978-3-86247-005-1.

Band 13:   Milena A. Dimova (2010)
Untersuchungen zur Epidemiologie von *Pythium aphanidermatum* in Abhängigkeit von den Umgebungsbedingungen bei der Gewächshausgurke (*Cucumis sativus* L.).
ISBN 978-3-86247-033-4.

Band 14:   Claudia Patricia Pérez-Rodríguez (2010)
Physiologische Veränderungen in Früchten der Solanaceaengewächse in Abhängigkeit von physikalischen Elicitoren während der Produktion und nach der Ernte.
ISBN 978-3-86247-066-2.

Band 15:   Charles Adarkwah (2010)
Integrated management of the stored-product pest insects *Corcyra cephalonica, Cadra cautella, Sitophilus zeamais* and *Tribolium castaneum* by use of the parasitic wasps *Habrobracon hebetor, Venturia canescens, Lariophagus distinguendus* and neem seed oil.
ISBN 978-3-86247-077-8.

Band 16:   Christoph von Studzinski (2010)
Angewandte Methoden der xenovegetativen Vermehrung.
ISBN 978-3-86247-088-4.

Band 17:   Tanja Mucha-Pelzer (2011)
Amorphe Silikate – Möglichkeiten des Einsatzes im Gartenbau zur physikalischen Schädlingsbekämpfung.
ISBN 978- 3-86247-106-5.

Band 18:   Diego Miranda (2011)
Effect of salt stress on physiological parameters of cape gooseberry, *Physalis peruviana* L.
ISBN 978- 3-86247-119-5

Band 19:   Franziska Beran (2011)
Host preference and aggregation behavior of the striped flea beetle, *Phyllotreta striolata*.
ISBN 978- 3-86247-188-1

Band 20:   Mohammed Abul Monjur Khan (2011)
Induced biochemical changes and gene expression in *Brassica oleracea* and *Arabidopsis thaliana* by drought stress and its consequences on resistance to aphids.
ISBN 978- 3-86247-203-1.

Band 21:   Sandra Lerche (2012)
Untersuchungen zur Anwendung, Praxiseinführung und molekularen Identifizierung von Stamm V24 des entomopathogenen Pilzes *Lecanicillium muscarium* (Petch) Zare & W. Gams.
ISBN 978- 3-86247-248-2.

Band 22:   Carsten Richter (2012)
           Entwicklung und Überprüfung eines gasdichten Küvettensystems für
           Experimente unter hochgradig kontrollierten Bedingungen mit
           Gaswechselmessungen.
           ISBN 978- 3-86247-271-0.

Band 23:   Aksana Grineva (2012)
           Influence of the two stored grain pest insects *Sitophilus granarius* and
           Oryzaephilus surinamensis on temperature, relative humidity, moisture
           content, and mould growth in stored triticale.
           ISBN 978- 3-86247-279-6.

Band 24:   Carmen Büttner & Christian Ulrichs (2012)
           Aktuelle Themen in Landwirtschaft und Gartenbau am Beispiel von Südtirol.
           ISBN 978- 3-86247-279-6.

Band 25:   Juliane Langer (2012)
           Molecular and epidemiological characterisation of Cherry leaf roll virus
           (CLRV).
           ISBN 978- 3-86247-279-6.

Band 26:   Franziska Rohr-Doucet (2012)
           AOP-Variabilität in *Arabidopsis thaliana*-Kreuzungslinien – Auswirkungen
           auf die Resistenz gegenüber verschieden spezialisierten Lepidopteren-Arten.
           ISBN 978- 3-86247-329-8.

Band 27:   Vanessa Hörmann (2012)
           Lignin als biologische Barriere gegen Schimmelpize in Innenräumen.
           ISBN 978- 3-86247-330-4.

Band 28:   Jacqueline Kurth (2013)
           Auswirkungen verschiedener Düngerzusammensetzungen auf den Ertrag bei
           Schnittrosen unter Berücksichtigung des Anbauverfahrens.
           ISBN 978- 3-86247-336-6.

Band 29:   Juliane Langer, Carmen Büttner & Christian Ulrichs (2014)
           Kolumbien – klimatische und politische Voraussetzungen für eine
           landwirtschaftliche Produktion.
           ISBN 978- 3-86247-430-1.

Band 30:   Heike Luisa Dieckmann (2014)
           Detection of the European mountain ash ringspot associated virus (EMARaV)
           in Sorbus aucuparia L. in several European contries.
           ISBN 978- 3-86247-441-7.

Band 31:   Rima Marion Baag (2014)
           Analyse von trans-Resveratrol in historischen Rebsorten der Weinanbaugebiete
           Sachsen und Saale-Unstrut.
           ISBN 978- 3-86247-488-2.

Band 32:   Ayesha Rahmann (2014)
           Study of the protective effects of nano-structured silica and plant derived
           biomolecules on nuclear polyhedrosis virus affected silkworm larvae at the
           behavioral and molecular level.
           ISBN 978- 3-86247-495-0.

Band 33:   Bettina Gramberg (2015)
           Weiterentwicklung eines elektrochemischen Biosensors zum Nachweis von
           Pflanzenviren und Insektiziden.
           ISBN 978- 3-86247-512-4.

Band 34:   Wilhelm van Husen (2015)
           Artspezifische Aufnahme und Verteilung von Cadmium bei indigenen
           afrikanischen Gemüsearten und daraus abzuleitende Ernährungsempfehlungen.
           ISBN 978- 3-86247-523-0.

Band 35:   Jenny Roßbach (2015)
           European mountain ash ringspot-associated viras (EMARaV): diversity and
           geographic distribution in Europe.
           ISBN 978- 3-86247-547-6.

Band 36:   Silke Steinmöller (2015)
           Risikominderung der Verbreitung von Quarantäneschadorganismen der
           Kartoffel durch hygienisierende Maßnahmen.
           ISBN 978- 3-86247-550-6.

Band 37:   Christin Siewert (2016)
           Genomic and functional analysis of species within the Acholeplasmataceae –
           Phytoplasmas and Acholeplasmas.
           ISBN 978- 3-86247-579-7.

Band 38:   Angela Köhler (2016)
           Untersuchungen zur Phenolglycosidkonzentration ausgewählter intra- und
           interspezifischer Kreuzungen salicinreicher Biomasseweiden.
           ISBN 978- 3-86247-581-0.

Band 39:   Nicolas Meyer (2016)
           Vergleichende ökophysiologische Untersuchung verschiedener Baumarten zur
           Verwendung als Straßenbegleitgrün in Berlin.
           ISBN 978- 3-86247-586-5.

Band 40:   Stefanie Schläger (2017)
           Identification of variation within sex pheromone blends of various Maruca
           vitrata populations for refining pheromone lures and traps in Asia.
           ISBN 978- 3-7369-9570-3.

Band 41:   Elisha Otieno Gogo (2017)
           Pre- and postharvest treatments for the quality assurance of African indigenous
           leafy vegetables.
           ISBN 978-3-7369-9650-2.

Band 42:   Luise Dierker (2017)
           Interaktion des RNA2-kodierten Transportproteins (MP) des *Cherry leaf roll
           virus* (CLRV) mit dem viralen Hüllprotein (CP) und pflanzlichen
           Wirtsfaktoren
           ISBN 978-3-7369-9670-0.

Band 43:   Nadja Förster (2017)
           Antikarzinogenes Potential ausgewählter Glucosinolate von *Moringa oleifera*
           ISBN 978-3-7369-9704-2.

Band 44:   Steffen Pallarz (2018)
             Data driven classification of host-plant response (virus-plant)
             ISBN 978-3-7369-9731-8.

Band 45:   Vanessa Hörmann (2018)
             Biofiltration of indoor pollutants by ornamental plants
             ISBN 978-3-7369-9815-5.

Band 46:   Allan Ndua Mweke (2018)
             Development of entomopathogenic fungi as biopesticides for the management
             of Cowpea Aphid, *Aphis craccivora* Koch
             ISBN 978-3-7369-9908-4.

www.ingramcontent.com/pod-product-compliance
Lightning Source LLC
Chambersburg PA
CBHW060310220326
41598CB00027B/4293